食尋。蘊味

～細味最樸實無華的家鄉菜

黃淑儀

SAVOUR
THE
homestyle
cooking

前言

以為輕輕鬆鬆的儲存了幾拾個食譜，就急不及待的催促編
輯出新書，恨不得把新食譜馬上公諸於世，好待與您分享！

誰知道當編輯問我要「主題」時，我楞住了！怎麼會忘了
每本烹飪書都應該有個主題的呢？

尤其是近年的烹飪書屢次獲獎，得到認同與歡迎！應該是
欣慰！是喜悅！
但，也是壓力！

呆了幾個禮拜，不能再躲了，我很認真的細看這次的菜式；
很有趣，原來都是拍攝「吾淑吾食」所到地方吃到或學到
的菜式，特別有地方色彩！加上溫哥華菜商鄧孀帶五邑鄉
味的芋頭糕，都讓人有親切的媽媽的味道！

這是我需要的，也是我強調餸菜雖然簡單，但也要有親切
的、溫馨的感覺。讓親人，讓朋友在毫無拘束，絕無困擾
的心情下用膳。
希望您也感受到家鄉的味道！
媽媽的味道！

PREFACE

I naively thought I could put together a new book just like that after collecting tens of new recipes. I simply couldn't wait to share them with my dear readers.

But when my editor asked me what the theme of this new book is, I was dumbfounded. How could I forget every cookbook was written around a theme?

Though I'm honoured and glad that my cookbooks have earned much recognition and have won many international awards in recent years, such accolades also put pressure on me.

So I decided to let the thoughts sink in and a few weeks have passed. Oh well, I couldn't keep pushing it back anymore and I took a good look at the recipes again. Interestingly, I collected these recipes when I was shooting the travel-and-cooking TV show Eating Well With Madam Wong. They are food that I either tasted or learnt to make while travelling around Southern China. They all exude a regional flair and I especially included the taro cake recipe from Mrs Tang, a greengrocer in Vancouver, BC, Canada. These are all home-style recipes of mom's cooking that we all find comfort in.

That's actually what I need from food and what I couldn't stress enough — food doesn't have to be complicated to warm the hearts of the eaters. Food is meant to be approachable and welcoming, so that your friends and family could enjoy at ease without worrying about table manners or using your best silverware.

I hope you'd enjoy the hometown flavours I found. Bring back the taste of mom's cooking.

目錄

APPETIZER SOUP DESSERT DELICACY

CONTENTS

RUSTIC

樸素沒花巧的菜式，
細味食物的原味，
最能打動人心，
亦最能令人回味。

誠如封底所寫：
媽媽在哪，家鄉就在那！
簡單易做的蒸蛋糕，
用汁撈飯的燜餸，
老火湯⋯⋯這些是媽媽的拿手菜，
是伴隨成長的心靈雞湯，
亦是快樂的美味記憶。

透過工作的認識、朋友的分享和我自己的挑選，
食譜內的菜式，
我細分為：餐前小吃、添飯小菜、滋潤靚湯和甜點，
希望您能煮給心愛的人享用，
尤其是您的媽媽。

SIMPLE

08 鮮果沙律

FRESH FRUIT SALAD
refer to p.119

材料

木瓜	半個（去皮，切片）
香蕉	1條（去皮，切片）
奇異果	1個（去皮，切片）
草莓	4顆（切片）
玻璃樽	1個（用大滾水燙過，消毒）

沙律醬

原味乳酪	半杯
蜂蜜	2湯匙
薄荷葉	2湯匙（剁碎）
橙皮茸	2湯匙

做法

1. 玻璃樽先放木瓜片，舀入適量沙律醬；放入香蕉片，再舀入適量沙律醬。

2. 跟着放入奇異果片，再舀入適量沙律醬；放入草莓片鋪面即成。

POINT

1. 近年流行玻璃樽沙律，我亦貪新鮮試着做，效果不錯啊！三五知己在家聚會，每人一樽，隨意又美味。

2. 也可以簡單點，調勻沙律醬，倒入水果內，輕輕拌勻即可享用。

3. 可選用你愛吃的鮮果，但要注意的，高水分的鮮果不宜採用。

10 涼拌魷魚絲

SHREDDED SQUID COLD APPETIZER

refer to p.120

材料

魷魚	1 條（汆水 30 秒，關火焗 7 分鐘，浸冰水，切圈）
中芹	1 棵（切度）
紅蘿蔔	半條（切絲）
木耳	40 克（浸軟，切絲，汆水）

汁料

鹽	1 茶匙
糖	1 湯匙
黃芥末粉	1 湯匙
檸檬汁	1 湯匙
麻油	1 茶匙

做法

1. 調勻汁料。
2. 全部材料拌勻，倒入汁料，拌勻即可享用。

POINT

可以用墨魚代替魷魚，但墨魚要切絲才烹調。

12 涼拌豬肝

PORK LIVER COLD APPETIZER
refer to p.121

材料

豬肝	1 塊約 400 克
	（用花椒 1 茶匙、指天椒碎 1 隻份量，
	加過面水浸半小時）
薑	3 片
葱	3 棵

汁料

芫茜	1 棵（只要莖部，切碎）
鹽	半茶匙
糖	2 茶匙
生抽	1 湯匙
香醋	1 湯匙
麻油	少許

做法

1. 把浸過水的豬肝和薑葱，放入水中煮滾，關火焗 15 分鐘，取出用布或廚紙索乾水分，切片。
2. 豬肝澆上汁料，即可享用。

14 口水雞

MOUTH-WATERING CHICKEN
refer to p.122

材料

雞	1 隻（用雞一半重量的粗鹽醃雞的內腔、外皮，內腔塞入毛巾，用毛巾或廚紙包好，放入雪櫃醃一夜）
新鮮粉皮	2 張（剪成粗條）
皮蛋	4 個（一開八）

汁料

麻油	3 湯匙
香醋	3 湯匙
生抽	7 湯匙
紅糖	3 湯匙
薑米	1 湯匙
蒜茸	1 湯匙
芫茜碎	1 湯匙
花椒辣油	1 湯匙（隨個人喜愛加減）

＊將以上所有材料拌勻

後下材料

花生碎	1 杯
芫茜碎	2 湯匙

做法

1. 取走雞內腔的毛巾，沖去粗鹽。用猛火蒸 25 分鐘，涼後起肉，撕成條。
2. 粉皮略汆水至透明。
3. 碟內放入粉皮、雞肉，皮蛋圍邊。
4. 倒下汁料，上桌前再灑上後下材料，食時拌勻即成。

POINT

1. 蒸雞汁不要棄去，可用來煲飯、炒菜，味道鮮美。
2. 新鮮粉皮不易購買，可以用乾粉皮代替。乾粉皮的處理法是，
   將乾粉皮用清水略浸軟，再汆水至透明即可食用。

16

鄧嬸芋頭糕

MRS TANG'S TARO CAKE

refer to p123

APPETIZER

材料 A

臘腸	2 條約 3 安士（蒸熟後切粒）
臘肉	1 小塊約 3 安士（蒸熟後切粒）
冬菇	4 朵（浸軟，切粒）
瑤柱絲	1 湯匙
紹酒	1 湯匙

材料 B

芋頭	半個約 2 磅（去皮，刨粗絲）
片糖	1/8 塊
鹽	1 茶匙
五香粉	1 茶匙
水	2 杯

材料 C

粘米粉	半杯
糯米粉	半杯

材料 D

炒香白芝麻	隨量（可根據你的喜好添減）

做法

1. 將 A 料爆香,盛起待用。

2. 將 B 料倒入爆香 A 料的鑊內,兜勻,加入(1)同兜炒。

3. 將 C 料混合後篩勻,倒入(2)內輕輕拌勻。

4. 把(3)倒入已抹油的盆中,猛火蒸 30 分鐘,灑上芝麻即成。

POINT

1. 臘腸、臘肉蒸熟待涼後,會較容易切粒。

2. 白芝麻用水沖洗後,潷去水分,放入冷鑊內,以中火炒至乾身,
   漸變微黃即可。

沙葛鮮蝦炸雲吞

DEEP-FRIED WONTONS WITH YAM BEAN AND SHRIMP FILLING

refer to p.124

材料

沙葛	半個（去皮，切細粒）
蝦仁	半磅（用鹽抓洗 2 次，索乾水分，切碎）
蠔油	1 湯匙
即食鱈魚絲	1 包約 30 克（切碎）
芫茜碎	1 湯匙
廣東雲吞皮	150 克
蛋黃	1 個（打散）

做法

1. 用油爆炒蝦仁 10 秒即撈起，餘下油爆炒沙葛，蝦仁回鑊，下蠔油 1 湯匙，最後加入即食鱈魚碎及芫茜碎，兜勻盛起成餡料。

2. 蛋液塗在雲吞皮四邊，放入適量餡料，包成個人喜歡的形狀。

3. 用中慢火炸至金黃即可。

19

POINT

1. 餡料要待涼才包入雲吞皮，否則雲吞皮會容易穿破。

2. 因餡料已熟，將雲吞皮炸至金黃即可享用。

3. 剛撈起的炸雲吞，宜放在廚紙吸去油分，吃時更清爽。

22 芋頭餅 DEEP-FRIED TARO BALLS
refer to p125

材料

芋頭	1/4 個約 300 克（去皮，刨絲）
珠豆	半杯（花生連皮洗淨，用白鑊烘乾）
糖	5 湯匙
麵粉	3 湯匙

做法

1. 用糖略醃芋頭絲和珠豆，灑下糯米粉，輕輕地把芋頭絲捏成小圓球。

2. 放入慢火熱油內炸熟即可。

去芋頭皮時要帶手套，否則可能會令皮膚敏感，痕癢不已。 POINT

24 紫菜墨魚塊

DEEP-FRIED CUTTLEFISH CAKE WITH NORI SEAWEED
refer to p.126

材料

韓式紫菜	12 小片
已調味墨魚滑	300 克（撻打至起膠）
肥豬肉粒	2 湯匙
馬蹄	4 粒（去皮，拍扁，剁碎）
芫茜	1 棵（莖切碎，葉留用）
芝麻	半杯

做法

1. 墨魚滑內加入馬蹄碎、肥豬肉粒及芫茜莖碎，撻打一會，加入芝麻拌勻。薄薄抹一層墨魚滑在紫菜上，將芫茜葉貼在墨魚滑上。

2. 燒 40℃低溫油，把有墨魚滑的一方朝下，邊炸邊轉動，確保受熱均勻，炸熟墨魚滑後才轉身，改大火炸 20 秒即把紫菜取出。

POINT　可放入一小塊墨魚滑試油溫，看見墨魚滑四周冒泡就可以；如油溫太高，會很容易炸焦。

STEAMED RADISH DUMPLINGS

蘿蔔丸子

refer to p127

材料

蘿蔔	1 個約 2 斤（去皮，刨絲，用 1 茶匙鹽略醃）
粘米粉	1.5 杯
糯米粉	3 湯匙
胡椒粉	1 茶匙
紹興酒	2 湯匙
芝麻	半杯（裝飾用）

配料

臘腸	1 條（蒸軟，切碎）
臘肉	1/4 條（蒸軟，切碎）
蝦米	1 湯匙（浸軟，切碎）
冬菇	2 朵（浸軟，切碎）
蒜頭	1 粒

27

做法

1. 芝麻用水沖洗後，瀝去水分，用白鑊烘乾至金黃色。

2. 擠去蘿蔔絲水分，加入已篩的粘米粉、糯米粉及胡椒粉，拌勻。

3. 熱鑊冷油，把配料炒香，灒酒，倒入（2）內拌勻，做成丸子放入蒸籠，
   猛火蒸 12 分鐘，再焗 3 分鐘，灑上芝麻即可享用。

POINT

1. 圓滾滾的蘿蔔不易刨絲，將它切開一半，會較容易掌握。

2. 用鹽醃蘿蔔絲的原因是，可迫出蘿蔔的苦澀味。

3. 炒配料時用熱鑊冷油，可讓材料隨油慢慢加熱，不易燒焦。

DUMPLINGS WITH
CHAYOTE PORK FILLING

refer to p.128

材料

佛手瓜	2 個約 550 克（去皮，刨絲，用 1 茶匙鹽醃一會，漌去水分）
免治豬肉	250 克
冬菇	2 朵（浸軟，切碎）
圓形餃子皮	半斤

豬肉調味料

鹽	半茶匙
糖	半茶匙
生粉	1 茶匙
麻油	1 茶匙
蛋白	1 個（蛋黃留作黏口用）

做法

1. 豬肉加調味料後撻打幾下，與佛手瓜絲及冬菇碎混合成餡料。

2. 餃子皮包入 1 茶匙餡料，用蛋黃塗邊，捏緊即可。

3. 煮沸一煲水，放入餃子，煮至餃子浮面，加凍水半杯，水再滾起，
   即可把餃子取出，用醋或生抽伴食。

1. 經鹽醃的佛手瓜絲會溢出水分，但拌餡前只需瀝去水就成，不
   用將它擠乾，否則享用時餡料會較乾，不夠 juicy！

2. 待餃子煮至浮面，再加入凍水半杯，是烹調餃子的竅門；這樣
   餃子餡一定會熟，而外皮不會煮糊了。

POINT

32 茶香雞翼

TEA-SCENTED CHICKEN WINGS
refer to p.129

材料

雞翼中段	2 磅
濃縮鹽水	3 杯（將 3 杯水煮溶半杯粗鹽）
薑	8 片
桂花醬	2 湯匙
桂花茶	半杯（桂花與紅茶葉焗成濃茶）

桂花醬

海鮮醬	8 湯匙
杜侯醬	4 湯匙
桂花糖	2 湯匙
磨豉醬	1 湯匙
花生醬	1 湯匙
腐乳	3 塊（搓爛）
片糖	半塊（搗碎）

做法

1. 桂花醬：將所有桂花醬的材料用中慢火邊煮邊攪勻，煮滾後關火。待涼後，舀進已用大滾水消毒的玻璃瓶內貯存。

2. 將濃縮鹽水煮滾，放入雞翼，待再滾起即關火焗 10 分鐘。取出雞翼，沖凍水半小時即成。

3. 熱鑊下油 1 湯匙，爆香薑片，加入雞翼、桂花醬兜勻後，倒入桂花茶，猛火兜炒至乾水即可。

POINT

1. 此菜亦可加入雞肝同煮，在每邊雞肝剠一刀，讓它快入味，易熟。

2. 詳細的「冰鮮雞翼處理法」，可參考我的另一本食譜《我的拿手菜》第 22 頁。

DELICACY

DELICACY

啤酒鴨

BRAISED DUCK
IN BEER SAUCE
refer to p.130

材料

鴨	1 隻（洗淨，切塊，汆水）
紅棗	8 粒

配料

八角	2 粒
薑	2 塊
蒜頭	4 粒
指天椒	1 隻
冰糖	2 粒

調味料

啤酒	2 罐
鹽	1 茶匙
老抽	1 湯匙

做法

1. 熱鑊冷油，爆香配料，放入鴨塊，再爆香。
2. 加入調味料和紅棗，燜 2 小時即可享用。

若用壓力煲，半小時即可享用。　　POINT

35

獨子蒜栗子燜雞

BRAISED CHICKEN WITH CHESTNUTS AND GARLIC
refer to p.131

材料

雞扒	4 個（切塊，醃過）
獨子蒜	12 粒（去衣）
去殼栗子	18 粒（汆水 10 分鐘，水可留作燜雞用）

配料

薑	2 片
紅椒	1 隻
八角	1 粒
冰糖	2 粒

醃雞料

鹽	半茶匙
糖	1 茶匙
生抽	1 湯匙
紹興酒	1 湯匙
麻油	1 茶匙
胡椒粉	少許

調味料

生抽	2 湯匙
蠔油	2 湯匙
糖	1 湯匙

做法

1. 鑊內放入配料爆香，放入雞塊爆炒，待雞皮溢出少許油後放入獨子蒜。
2. 待獨子蒜爆透後放入栗子，加入調味，可略加少許焓栗子水，約燜 3 分鐘即可上碟。

POINT

1. 栗子去殼、去衣都比較麻煩，有指可切去頂部粗枕，放入微波爐「叮」2 至 3 分鐘，即能殼起衣離。
2. 栗子燜雞可變化為栗子燜鴨，做法一樣，但水要蓋過鴨面燜 45 分鐘才能上碟。

DELICACY

38 碌雞

ROLLING BRAISED CHICKEN
refer to p.132

材料

雞	1隻（洗淨，抹乾，用1茶匙鹽抹勻內腔）

調味料

紹酒	半杯
美極鮮醬油	1/3 杯
片糖	2/3 片
薑	4 片
葱	2 條
滾水	1/2 杯

做法

1. 熱鑊下 2 湯匙油，把雞煎碌 5 分鐘，讓雞皮顏色金黃。
2. 倒入調味料，收中火，雞胸向下，加蓋煮 5 分鐘，揭蓋後要不斷將雞碌滾，大約 20 分鐘直到汁液濃稠即可。

POINT

用這方法烹調，可令雞皮的顏色平均、漂亮。

40 甜椒雞

ROMAN-STYLE CHICKEN STEW

refer to p.133

材料

雞柳	6 條（用鹽、黑椒粉各 1 茶匙略醃）
大洋葱	1 個（切幼絲）
番茄	4 個（切塊）
三色甜椒	各半個（去籽，切塊）
水瓜鈕	1 湯匙（舂碎）
白酒	3/4 杯
海鹽	1 茶匙
黑椒碎	2 茶匙
橄欖油	4 湯匙

做法

1. 洋葱絲放入冷鑊冷油內，用中火爆炒至軟身，放入雞塊，爆炒約 10 分鐘，倒入白酒再煮 5 分鐘，取出雞塊待用。

2. 此時洋葱已煮爛，加入番茄、甜椒、水瓜鈕，下鹽、黑椒碎調味，蓋着煮約 20 分鐘至腍，雞肉回鑊煮熱即可上碟。

POINT

這菜式是我去年夏天往意大利的鄉下學的，是意大利媽媽們的家常菜。沒有花巧，外觀樸實，但味道一流。

DELICACY

DELICACY

南瓜雞盅

STEAMED PUMPKIN STUFFED WITH CHICKEN AND ONION

refer to p.134

材料

日本南瓜	1 個約 2 千克
	（在 1/4 處切去頂部，去籽）
雞扒	3 塊
	（切塊，用少許鹽、胡椒粉作底味略醃）
洋葱	半個（切碎）
桂花醬	2 湯匙
葱絲	1 條份量

做法

1. 南瓜頂部、南瓜心用少許桂花醬抹勻。

2. 用油爆香洋葱碎，加入雞肉，倒入餘下的桂花醬，爆熟雞肉即可放
   入南瓜內，蓋上南瓜頂部。

3. 把南瓜放在湯碗內，隔水蒸 20 分鐘。

4. 灑上葱絲，灒熟油，即可上桌

POINT

1. 桂花醬做法看第 32 頁的茶香雞翼。

2. 南瓜放入湯碗內，可固定南瓜的形態，因蒸熟的南瓜會較軟。

煎
蟹
餅

CRAB CAKE
refer to p.135

材料

蟹	1 隻（蒸 10 分鐘，拆肉）
洋葱	半個（切粒）
乾葱頭	2 粒（切碎）
煙肉	4 片（切碎）
蘑菇	6 粒（切碎）
豌豆	半杯
馬蹄	4 粒（拍碎）
沙律醬	2 湯匙
芝士碎	1 湯匙
蛋	3 個（打勻）

做法

1. 熱鑊下油,用中火爆香洋葱、乾葱頭,加入煙肉、蘑菇、豌豆、馬蹄及蟹肉,拌勻,盛起。

2. 倒入沙律醬及芝士碎,拌勻,最後加入蛋液,拌勻。

3. 舀出約 1 湯匙分量,慢火煎香即可享用。

POINT

1. 如菜式需要保持蟹身完整,可放蟹入冰格,避免牠掙扎時弄甩爪螯。

2. 宜用牙刷刷去蟹身的污糟物才烹調。

滬江明蝦

FRIED PRAWNS IN OSMANTHUS AND DISTILLERS GRAIN SAUCE

refer to p.136

材料

中蝦	900克（去殼，挑腸，在背部剅一刀，用鹽抓洗2次，索乾水分）
沙葛	半個（去皮，切粗條，墊底用）

配料

冬菇	2朵（切碎）
薑	4片（剁成薑米）
蒜頭	4粒（剁成蒜茸）
葱	2條（切成葱花）

醬汁

桂花醬	2湯匙
酒釀	2湯匙
生抽	1湯匙
陳醋	1茶匙
糖	1湯匙
麻油	少許

做法

1. 燒略多油，將蝦半煎炸至僅熟，撈起。
2. 倒去過多油分，剩下少許油，用中火爆香配料，潷酒，下醬汁煮沸，蝦回鑊，兜炒至每一隻蝦都沾滿醬汁，盛起鋪在沙葛上；同食可降燥熱。

POINT

1. 宜挑選少橫紋、少根的沙葛，肉質較脆嫩。
2. 不同於芋頭、番薯等根莖類，將沙葛去皮是不用刀或刨，而是用手撕去皮。
3. 沙葛可生吃，味道清甜，口感脆嫩。
4. 桂花醬做法看第32頁的茶香雞翼。

50

豉椒紫蘇炒蜆

STIR-FRIED CLAMS WITH CHILLIES AND SHISO

refer to p.137

材料

蜆	1 斤（氽水，一張口即撈起）

配料		調味料	
薑	2 片（剁成薑米）	生抽	1 湯匙
蒜頭	4 粒（剁成蒜茸）	蠔油	1 湯匙
紅椒	1 隻（切碎）	糖	1 茶匙
豆豉	2 湯匙（用水沖洗）	水	少許
紫蘇葉	4 片（切碎）		

做法

1. 熱鑊下油，順序爆香配料，加入蜆，兜勻。
2. 加入調味料，兜勻，即可享用。

POINT

1. 凡用來炒的貝殼類都需要先氽水，待殼張開，才下鑊兜炒，這樣可確保貝類新鮮；而且先氽水再兜炒的方法，可保持肉質鮮嫩。

2. 紫蘇有解毒袪腥的作用，所以多配上貝類、蟹同烹調。

砂鍋酒香多寶魚

WINE-SCENTED TURBOT FILLET IN CLAY POT
refer to p.138

材料

多寶魚	1 條（約斤半，洗淨去鱗）
紹興酒	4 湯匙
紅椒	2 隻（斜刀切塊）

醃料

鹽	1 茶匙
胡椒粉	少許
薑汁	1 茶匙
葱汁液	少許
	（葱白用刀背拍碎，取出的汁液）
生粉	1 茶匙
生油	1 湯匙

配料

老薑	5 片
紅葱頭	8 粒
蒜頭	8 粒
葱	5 棵（切 2 吋長）

DELICACY

做法

1. 魚起肉，切厚片，用醃料略醃。

2. 用中火炒香配料，放入砂鍋內，倒入魚片，蓋上砂鍋蓋，在蓋邊淋下紹興酒 4 湯匙，轉小火焗 4 分鐘至魚熟，灑上紅椒即可。

POINT

1. 起魚脊肉時，刀要緊貼魚骨。

2. 魚骨可用醃料略醃，煎香後下酒，非常美味。

STEAMED SHIITAKE MUSHROOM STUFFED WITH MINCED SHRIMP FILLING

refer to p.139

材料

冬菇	12 朵（大小要統一，處理好）
蝦	250 克（去腸，抓洗兩次，索乾拍碎）
肥肉	1 小塊（剁碎，約 1 湯匙）
馬蹄	3 粒（去皮，拍碎，再剁細）
時菜	200 克（洗淨）
火腿末	1 湯匙（裝飾用）
蠔油	1 湯匙

醃料

蛋白	1 湯匙
胡椒粉	少許
麻油	少許

DELICACY

做法

1. 蝦肉醃味後撻打幾下，與馬蹄及肥肉混合後釀在冬菇內（冬菇內需索乾及塗上生粉），灑上火腿末，猛火蒸5分鐘即可。

2. 煲滾少許水，加入1湯匙油、1茶匙鹽，放入時菜冚蓋，滾2分鐘即可取出，伴在冬菇旁，淋上蠔油即可。

POINT

1. 冬菇處理法：將已浸軟洗淨的冬菇放入碗內，倒入已過濾的浸菇水（要浸過菇面，如水不足，須加添熱水），下2片薑、蔥2棵和1/4片片糖，用保鮮紙封好，再刺兩個小孔，放入微波爐「叮」10分鐘兩次。用油爆香2粒蒜頭，下冬菇，炒勻，加入草菇老抽2湯匙，倒入浸菇水，煮至水分收乾，下蠔油2湯匙炒勻，關火，待涼。將冬菇分數份裝在保鮮盒或保鮮膠袋內，放入冰格冷藏；烹調前不用解凍，可直接放入鍋內煮。

2. 冬菇內一定要索乾水分，以便容易塗上生粉。

58 豉椒蝦

FRIED PRAWNS WITH CHILLI BLACK BEAN SAUCE
refer to p.140

材料

中蝦	1斤（剪去鬚，挑去腸）
尖辣椒	4隻（洗淨，切斜刀）

配料		調味料	
豆豉	2湯匙	生抽	1湯匙
蒜茸	1湯匙	蠔油	1湯匙
薑米	1湯匙	糖	2茶匙
糖	1茶匙		
油	1湯匙		

做法

1. 用毛巾或廚紙吸乾中蝦水分，泡油待用。
2. 配料混合。
3. 用中火爆香配料，加入尖辣椒兜炒，中蝦回鑊兜炒，最後加進調味料炒勻，即可上碟。

 POINT　一定要吸乾中蝦的水分才泡油，否則泡油時熱油四濺，會容易受傷。

DELICACY

60 酸辣魚片

POACHED SLICED GRASS CARP IN SOUR SPICY BROTH

refer to p.141

材料

鯇魚脊肉	1塊（切片，醃半小時）
薑	4片（切絲）
酸筍絲	半杯
紅椒	1隻（切碎）
辣椒醬	1湯匙（按個人喜好添減）
滾水	約 1 1/2 杯
糖	1茶匙
芫茜	1棵（切碎）

醃料

鹽	半茶匙
生粉	1湯匙
糖	1/3 茶匙
蛋白	1個
胡椒粉	少許

做法

1. 燒熱鑊下2湯匙油，爆香薑絲、酸筍絲、紅椒碎、辣椒醬，倒入過面滾水。

2. 加入1茶匙糖，放入已經醃過的魚片，散開， 一滾即熄火，芫茜灑面即可享用。

POINT

用蛋白醃魚肉，可令魚肉更加嫩滑。

DELICACY

菠菜魚柳卷

SPINACH FISH ROLLS

refer to p.142

材料

桂花魚	1 條（起肉切片，以鹽及胡椒粉略醃）
菠菜	1 斤（洗淨，分開葉、莖）
紹酒	1 湯匙
葱	3 棵（汆水）
水	3 杯

汁料

牛奶	半杯
松子	2 湯匙（白鑊烘香）
鹽	半茶匙
胡椒粉	少許

做法

1. 魚頭、尾、骨加薑片煎香，潛酒，潛水，煮約 10 分鐘至出味，撈起；魚湯留用。

2. 菠菜莖切度，用魚片捲起，用葱紮實，放入魚湯中煮約 2 分鐘，取出。

3. 菠菜葉放入魚湯中，一焯即撈起，放入攪拌機內，加汁料打成濃汁，淋在魚卷上即可。

POINT

1. 宜購買大條的桂花魚，魚肉較厚和結實。

2. 要用熱鑊冷油煎魚；煎魚的首要秘訣是耐性，要先將一邊煎至金黃，才翻轉煎另一邊，否則翻來覆去，魚身會「甩皮甩骨」。

3. 可以用灼軟的韭菜花代替葱紮實魚卷。

PAN-FRIED PORK PATTIES
WITH LOTUS ROOT
AND DRIED TANGERINE PEEL
refer to p.143

材料

蓮藕	1小節約半斤（刮去皮，切極細粒）
免治豬肉	半斤（用醃料先調味）
陳皮	1角（浸軟，剁極幼）

醃料

蛋液	半個
鹽	1茶匙
糖	半茶匙
紹興酒	1茶匙
胡椒粉	少許
薑茸	少許
芫茜碎	少許
葱花	少許

做法

1. 蓮藕粒倒入滾水中汆水一會，即撈起，再汆凍水，撈起後
   瀝去水分，備用。

2. 把蓮藕粒及陳皮粒拌入已醃好的免治豬肉內；手抹油，先
   將豬肉料搓成丸子，再壓扁。

3. 以中火煎熟陳皮豬肉蓮藕餅即可。

這是一個可以令你添飯的伴飯菜；也可以搖身變為小點，用
荷葉餅夾着吃，蘸點喼汁，非常美味。

68 山藥炒羊肉

STIR-FRIED SLICED LAMB
WITH FRESH YAM
refer to p.144

材料

羊肉	12 兩（切片）
山藥	半斤（切片，汆水 20 秒去黏液）
中芹	1 棵（去葉，切度）
薑	4 片
蒜頭	4 粒（略拍）
葱	2 棵（切度）
紅椒	2 隻（切碎）

調味料

鹽	1 茶匙
生抽	1 湯匙
酒	2 湯匙
胡椒粉	少許
麻油	少許

做法

1. 熱鑊涼油，只用少許油煸炒羊肉至出水。潷去水分，盛起。
2. 爆香薑、葱、蒜和紅椒，倒入山藥、中芹及羊肉，加入調味料兜勻即可。

POINT

1. 刨山藥皮時，宜戴上手套，因它的汁液會令皮膚敏感。
2. 已刨皮的山藥應浸在鹽水中，可防止山藥因氧化而變成灰黑色。

DELICACY

70 釀苦瓜

BITTER MELON STUFFED
WITH GROUND PORK FILLING
refer to p.145

材料

長形苦瓜	1 條（切環形，約 1 吋厚，挖去籽）
免治豬肉	120 克（略醃）
豆豉	1 湯匙（沖乾淨，與蒜茸、陳皮茸、糖各 1 茶匙調勻）

豬肉醃料

薑汁酒	1 湯匙（把 1 湯匙薑茸，擠出汁液與 1 湯匙紹興酒混合）
生抽	1 茶匙
生粉	1 茶匙
麻油	1 茶匙
胡椒粉	少許

汁料

鹽	半茶匙
糖	半茶匙
蠔油	1 湯匙
水	2 湯匙

做法

1. 煲內注入 2 吋水，煮滾，加入油、鹽各 1 茶匙。放入苦瓜煮 2 分鐘，撈起沖凍水約 1 分鐘，取出，瀝乾水分待用。

2. 把醃好的免治豬肉釀入抹上生粉的苦瓜環內，用油煎至兩面熟（約 2 分鐘）。

3. 用另一個鑊，以 2 湯匙油爆香豆豉，加入汁料，煮熱；將已煮熱的汁料澆在苦瓜上，煮勻即成。

DELICACY

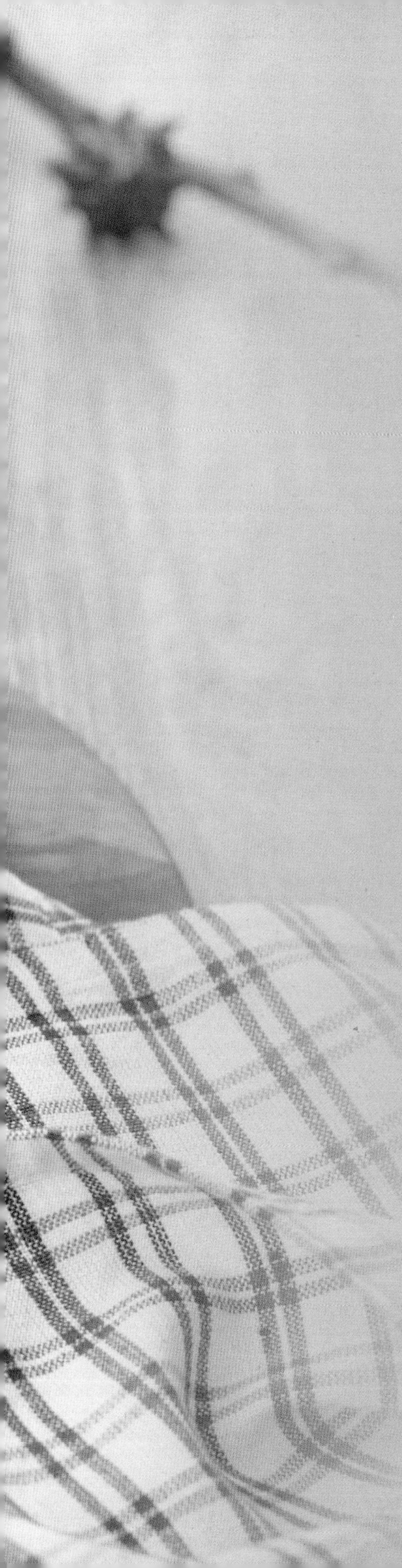

SPICY BRAISED
PORK TROTTER
refer to p146

香辣豬手

材料
豬手	1隻

（斬件，加入香料包氽水1小時，取出沖凍水）

氽水香料
桂皮	1塊
八角	2粒
香葉	3片
甘草	3片

* 放入紗袋內

配料
薑	2片
蒜頭	4粒
花椒	1湯匙
指天椒	1隻

調味料
紹興酒	1湯匙
片糖	半塊
老抽	1湯匙
鹽	半茶匙
滾水	1杯

做法

1. 熱鑊下油，爆香配料，加入已汆水的豬手，豬手爆透後加入調味料。
2. 燜 45 分鐘即可或用高壓煲煮 15 分鐘。

POINT　如香料不放入紗袋內，品嚐時可能會咬到香料，而不能盡情享用。

DEEP-FRIED CHITTERLINGS WITH SWEET AND SOUR SAUCE

糖醋豬大腸

refer to p147

材料

冰鮮豬大腸	2條（用4片薑、葱2棵出水1小時，切片，剪去肥膏）

配料

薑	2片
葱	2棵
蒜頭	2粒

豬大腸漿粉

生粉	8湯匙
麵粉	1茶匙
蛋白	2個
凍水	1/4杯

* 將凍水逐少倒入其他漿粉料中拌勻

糖醋料

鹽	半茶匙
蒸魚豉油	1湯匙
生粉	半茶匙
糖	1湯匙
鎮江醋	1湯匙
水	2湯匙

75

做法

1. 把處理好的大腸放入漿粉中拌勻。

2. 預備一鍋油，慢火先炸一次，撈起。

3. 油再加至大熱，把大腸再炸一次。

4. 爆香薑、葱、蒜，把糖醋料煮滾，大腸回鑊兜勻即可。

POINT

可以用新鮮豬大腸代替冰鮮的。處理方法：用醋及鹽抓洗豬大腸數次，反轉豬大腸，剪去肥膏，同樣用醋及鹽抓洗幾次；反轉還原豬大腸後再抓洗。煮沸一鍋水，放入豬大腸汆水1分鐘，取出，再以薑、葱同煮1小時。取出切小段，待用。

78

蝦米燴白菜

BRAISED BOK CHOY
WITH DRIED SHRIMPS

refer to p.148

材料

白菜	1斤（洗淨）
薑米	1湯匙
蝦米	1湯匙（洗淨）

做法

1. 熱鑊下油爆香蝦米和薑米，沖入1杯滾水，煮約3分鐘，讓蝦米出味。

2. 放入白菜，煮約2分鐘即可！

POINT

1. 蝦米不需要浸軟，只需要洗淨即可，否到味道會變淡。

2. 可用「口水雞」的蒸雞汁煮白菜，好味到不得了。

80 豆腐素餅

VEGETARIAN TOFU PANCAKES
refer to p.149

材料

洋葱	半個（切碎）
蒜茸	1 湯匙
馬蹄	4 粒（去皮拍碎）
冬菇	2 朵（切碎）
鮮百合	1 個（切碎）
合桃	1 湯匙
花生	1 湯匙
芝麻	2 湯匙
蒸豆腐	2 塊（捏碎）

三種果仁同壓碎

調味料

生粉	1 湯匙
鹽	半茶匙
胡椒粉	1 茶匙

做法

1. 用油爆香洋葱及蒜茸，再順序爆炒馬蹄、冬菇及鮮百合。

2. 果仁碎倒入（1）內，拌勻，再放入豆腐，下調味料，拌勻，再捏成小圓餅。

3. 熱鑊下少許油，慢火煎香豆腐素餅即成。

POINT

可以伴泰式甜酸醬享用。

DELICACY

STIR-FRIED MUNG BEAN VERMICELLI WITH WOOD EAR AND DRIED TOFU

refer to p150

木耳豆乾炒粉絲

材料		配料	
粉絲	2 扎	乾葱	2 粒（切絲）
		木耳絲	15 克（浸軟後切絲）
粉絲調味料		菜脯	30 克（切絲）
鹽	1 茶匙	五香豆乾	1 塊（切絲）
油	1 湯匙	紅蘿蔔	半條（刨絲）
胡椒粉	少許	荷蘭豆	50 克（切絲）
老抽	1 湯匙	中芹	1 棵（切絲）
		紅椒	1 隻（切絲）

做法

1. 燒滾一鍋水，放入粉絲調味料，放入粉絲，汆至軟化即撈起；
   粉絲用剪刀剪開一半。
2. 順序爆炒配料，加鹽半茶匙，加入粉絲快手兜炒，兜勻所有配
   料和粉絲即可。兜炒時如覺得太乾，可酌量加少許上湯。

POINT

1. 可用汆水代替將粉絲浸軟才煮，粉絲會更加入味，同時質
   感會煙韌些，炒時不易斷。
2. 炒粉絲成功的首要條件是乾濕度要適中，所以如覺得乾，
   要加些上湯。

烏雞石斛湯

BLACK SILKIE CHICKEN SOUP WITH SHI HU
refer to p.151

材料

烏雞	1 隻（洗淨，汆水）
石斛	20 克
花旗參	20 克
海竹頭或玉竹	40 克
土茯苓	40 克（鮮土茯苓則用 160 克）
無花果	4 個
乾螺頭	80 克
腰果	120 克
薑	5 片
水	20 杯

海竹頭

做法

1. 乾螺頭的處理方法：乾螺頭洗淨後，浸過夜，倒入電飯煲中，加 1 粒冰糖，按掣煮至乾水，就可採用。
2. 全部材料放入煲內，用大火煲滾後再煲 10 分鐘，轉中慢火煲 3 小時即可。
3. 隨個人口味下鹽調味。

POINT

1. 經處理的乾螺頭冷卻後可分裝膠袋，置冰格，隨時取出烹調。
2. 海竹頭有清熱、養陰、生津、潤燥、潤肺的作用。
3. 這湯有養陰、清熱、生津、益胃的功效；適宜經常捱夜、口乾舌燥、夜不安眠、糖尿病患者及高血壓病患享用。

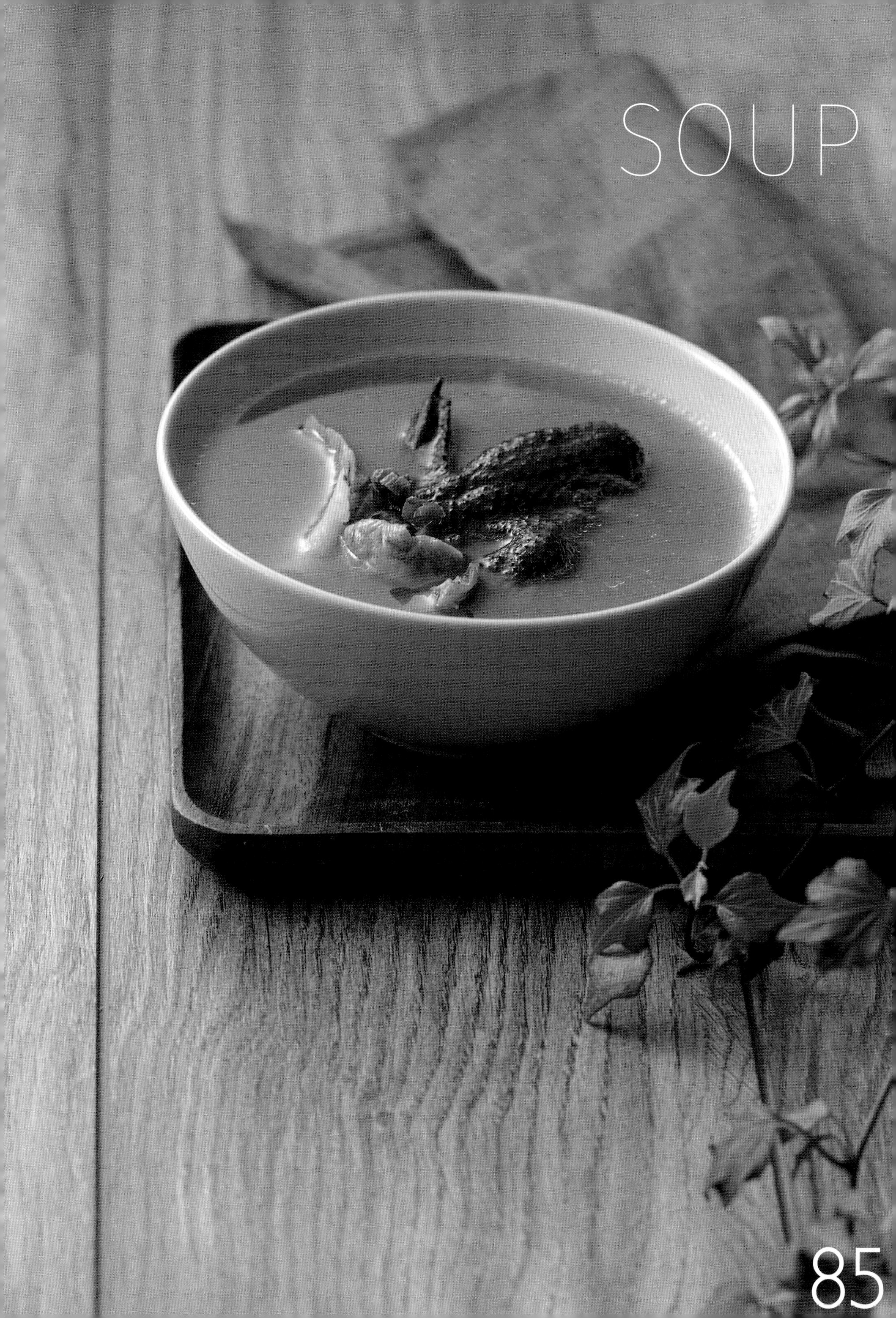
SOUP

86 南瓜鯽魚湯

CRUCIAN CARP SOUP
WITH PUMPKIN

refer to p.152

材料

鯽魚	1 條（洗淨後用油煎香）
南瓜	半個約斤半（去皮、去籽、切塊）
西施骨	300 克（汆水）
無花果	2 粒
蜜棗	2 粒
薑	2 片
水	18 杯

魚袋

做法

1. 鯽魚放入魚袋內。
2. 所有材料放入煲內，用大火煲滾後再煲 10 分鐘，然後改慢火再煲 3 小時即可。
3. 隨個人口味下鹽調味。

POINT

1. 通常煲魚湯都應該把魚放入魚袋，以免魚骨留在湯中。
2. 無論煎魚或汆水，加兩片薑可辟腥味。
3. 鯽魚有祛濕、滋陰、行氣、活血的功效。

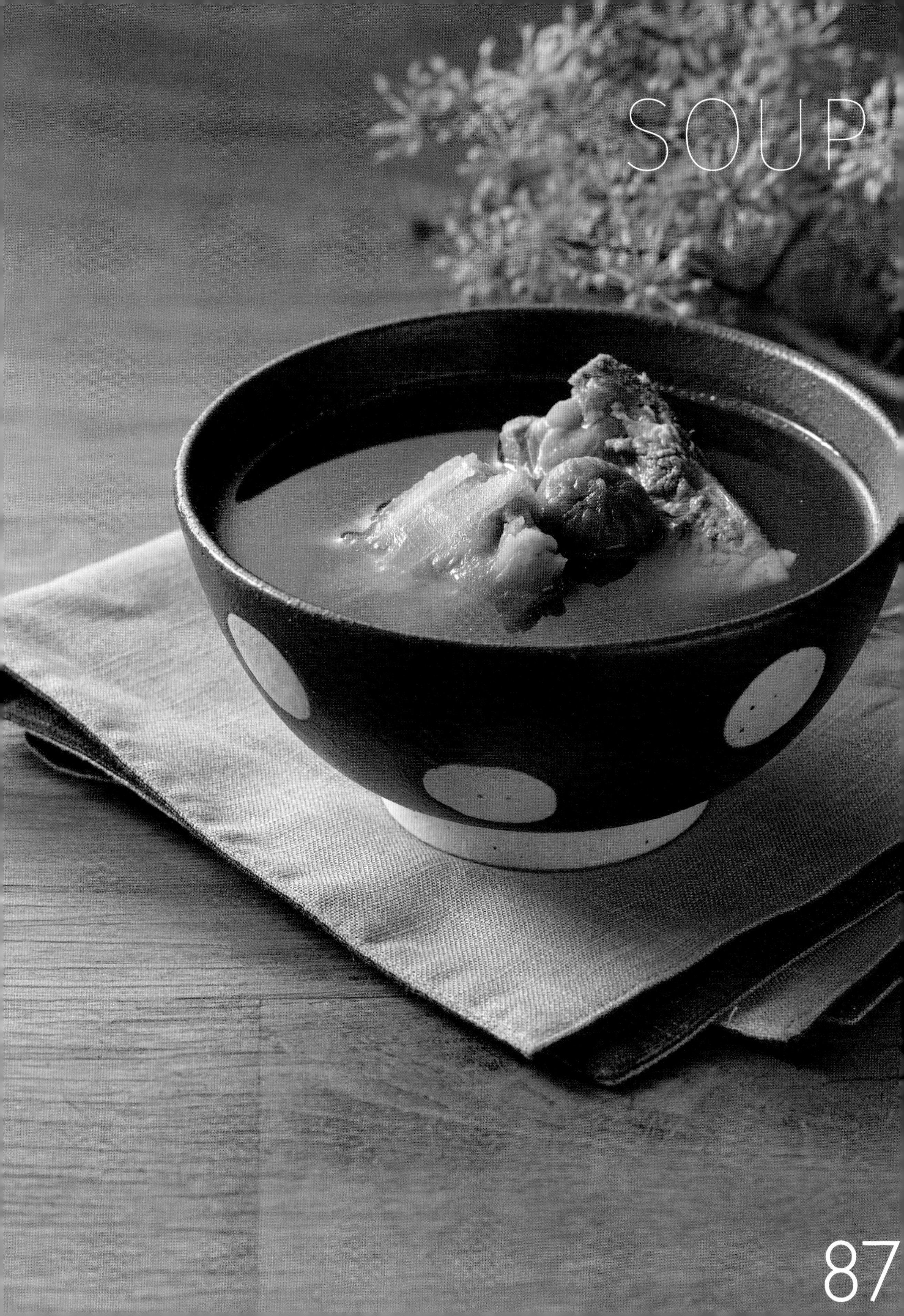
SOUP

SOUP

BEEF SOUP WITH BUTTON MUSHROOMS

refer to p153

蘑菇牛肉湯

材料

牛肉	1 磅（切丁）
牛油	1 湯匙
水	6 杯

調味料

百里香	1/4 茶匙
白酒	半杯
茄膏	2 湯匙

配料

蘑菇	半磅（切丁）
洋蔥	1 個（切粒）
紅蘿蔔	半個（去皮，切粒）
西芹	2 枝（切粒）
薯仔	1 個（去皮，切粒）

做法

1. 用牛油爆炒牛肉粒，加入調味料略兜炒，盛起。

2. 用另一個鑊爆炒配料（除了薯仔）3 分鐘，加入（1）及水煮 50 分鐘。

3. 最後加入薯仔粒，再滾 10 分鐘試味，隨個人口味下鹽調味。

90

魚頭黨參湯

FISH HEAD SOUP
WITH DANG SHEN

refer to p.154

材料

大魚頭	1 個（劈開兩邊）
豬骨	300 克（汆水）
黨參	40 克
北芪	40 克
天麻	30 克
馬蹄	10 粒（去皮）
紅棗	10 粒（去核）
薑	2 片
滾水	14 杯
米酒	4 湯匙（分 2 次用）

做法

1. 把魚頭煎至兩面金黃，先潷 2 湯匙米酒，倒入滾水 2 杯。

2. 把（1）轉到煲內，倒入其餘水及材料，慢火煲 2 小時。關火，倒入餘下 2 湯匙米酒即可。

3. 隨個人口味下鹽調味。

 POINT　這湯有補腦、養血、袪風的療效。

92

西施骨竹筍湯

PORK SOUP
WITH BAMBOO SHOOTS
refer to p.155

材料

鮮竹筍	2 個（去筍衣，切塊）
西施骨	600 克（汆水）
黃豆	300 克（洗淨，略浸）
潮州鹹酸菜	1 個（洗淨，切片）
乾墨魚	150 克（洗淨，去膜）
陳皮	1 個（浸軟）
蜜棗	3 粒
薑	5 片
水	18 杯

做法

1. 所有材料放入煲內，用大火煲滾後再煲 5 分鐘，收中慢火再煲 2 小時即可。
2. 隨個人口味下鹽調味。

POINT

處理竹筍是非常容易的。首先在竹筍身直切一刀，逐片剝去竹殼，待見到筍肉就成；宜切去竹筍底部已呈纖維化的部分。

SOUP

黨參北芪牛尾湯

OXTAIL SOUP
WITH DANG SHEN
AND BEI QI

refer to p.156

材料

牛尾	1 條（汆水）
黨參	20 克
北芪	20 克
巴戟	20 克
杜仲	20 克
土茯苓	30 克
製芡實	30 克
草果	3 個
香葉	3 片
陳皮	1 個（浸軟）
蜜棗	5 粒
薑	5 片
水	20 杯

製芡實

做法

1. 全部材料放入煲內，用大火煲滾後再煲 5 分鐘，再收中慢火煲 3 小時即可。
2. 隨個人口味下鹽調味。

 POINT   這湯有堅腎、固腎、益精的功效。

SOUP

川貝母

LONGLI LEAF AND CHUAN BEI MU SOUP

refer to p157

龍脷葉川貝母湯

材料

瘦肉	600 克（汆水）
龍脷葉	30 克
川貝母	30 克
南北杏	30 克
百合	30 克
海竹頭或玉竹	30 克
蓮子	30 克
無花果	30 克
蠔豉	60 克
陳皮	1 個（浸軟，去瓤）
薑	5 片
水	20 杯

龍脷葉

做法

1. 所有材料放入煲內，用大火煲滾後再煲 5 分鐘，改中慢火煲 3 小時即可。
2. 隨個人口味下鹽調味。

POINT

1. 龍脷葉在菜市場內的生草藥檔有售。
2. 川貝母有清熱化痰、潤肺止咳的功效。
3. 這湯有清肺燥、止渴化痰的功效。

97

南瓜雜菌豆腐羹

CREAM OF PUMPKIN SOUP WITH ASSORTED MUSHROOMS AND TOFU

refer to p.158

材料

南瓜	半個（去皮，切片）
鮮冬菇	4 朵
蘑菇	6 粒
本菇	1 包
茶樹菇	1 包
嫩豆腐	1 塊（切碎）
淡奶	1 湯匙
水	4 杯
鹽	1 茶匙

鮮冬菇、蘑菇、本菇、茶樹菇 — 抹淨，切碎，汆水後備用

做法

1. 用猛火蒸南瓜片 8 分鐘，取出，加水半杯，用攪拌機打成杰糊。
2. 煲滾 4 杯水，加入南瓜糊，煮沸後加入豆腐及已汆水的雜菌。
3. 關火後加入淡奶，下鹽調味，拌勻即可盛起享用。

SOUP

100 乳酪香橙凍

ORANGE FROZEN YOGHURT CUP

refer to p.159

材料

橙汁	半杯
原味乳酪	1.5 杯
煉奶	1/3 杯
檸檬汁	2 湯匙
雲尼拿油	1 茶匙
鹽	1 撮

做法

將以上材料混合拌勻，倒入小杯內，置冰
格 2 小時即可享用。

POINT　這乳酪香橙凍表面看來平平無奇，但乳酪透出鮮橙香，好清新。

紅蘿蔔蛋糕

CARROT CAKE
refer to p.160

材料 A

蛋	4 個
菜油	1 杯
糖	1 杯
雲尼拿香油	1 茶匙

材料 B

麵粉	2 杯
發粉	2 茶匙
蘇打粉	1 茶匙
鹽	半茶匙
肉桂粉	2 茶匙

材料 C

紅蘿蔔	1 條（去皮，刨絲約 2 杯）

做法

1. 材料 A 混合，以中速打勻。

2. 材料 B 混合篩勻，分三次倒入（1）內，用慢速打勻，最後加入材料 C。

3. 放入已預熱 175℃（350 ℉）焗爐內焗 50 分鐘，取出後要靜置 10 分鐘即可享用。

POINT

如在加拿大，我會用 Dream Whip 1 包、牛奶半杯快速打 2 分鐘成忌廉伴紅蘿蔔蛋糕享用；也可伴已打發淡忌廉，亦很美味。

蒸焦糖蛋糕

STEAMED CARAMEL CAKE
refer to p.161

材料 A
片糖	2 片（剁碎）
滾水	半杯

材料 B
蛋	4 個
生油	半杯
蜂蜜	2 湯匙

材料 C
麵粉	1 杯
發粉	2 茶匙
蘇打粉	1 茶匙

做法

1. 將材料 A 的片糖碎放入小煲內，倒入滾水 1 湯匙，用慢火煮溶成金黃色的糖水，倒入餘下滾水，邊倒邊攪拌，待涼。

2. 先將材料 B 的雞蛋用中速打散，順序加入生油、蜂蜜及已冷卻的糖水，打至完全混合。

3. 把材料 C 一起篩勻，分三次加入（2）內，用慢速打勻，倒入焗盆內，蓋上保鮮紙 1 小時。

4. 隔水猛火蒸半小時即可。

POINT

1. 用片糖的原因是取其焦糖香味及色澤。

2. 麵糊倒入焗盆內，再蓋上保鮮紙 1 小時的原因是讓麵糊充分發酵，蒸後的蛋糕才會更鬆軟。

108 椰子蛋糕

COCONUT POUND CAKE
refer to p.162

材料 A

牛油	半磅（室溫）
忌廉芝士	1/3 塊（室溫）
糖	半杯
蛋	3 個（室溫）

材料 B

麵粉	1 杯
發粉	1 茶匙
蘇打粉	半茶匙
鹽	1 撮

材料 C

椰絲	半杯

做法

1. 將材料 A 內的牛油、忌廉芝士、糖逐樣加入盆內，用高速打勻；蛋要逐個加入，拌至完全融合後才加入另一個蛋。

2. 材料 B 混合篩勻，分三次倒入（1）內，轉慢速讓全部材料混合，最後加入 C 拌勻。

3. 放入已預熱 175℃（350 ℉）焗爐內焗 50-60 分鐘。

POINT

1. 可把部分椰絲灑在蛋糕上，不過要鋪錫紙防止「搶火」。
2. 亦可用糖薑代替椰絲，做法一樣，但味道完全不同。

日式梳乎厘芝士蛋糕

JAPANESE SOUFFLÉ CHEESECAKE
refer to p.163

材料 A

忌廉芝士	250 克（室溫）
牛油溶液	40 克
蛋黃	6 個（室溫）
餅粉	3 湯匙

材料 B

淡忌廉	4 湯匙
原味乳酪	2 湯匙
牛奶	2 湯匙
檸檬汁	1.5 湯匙

材料 C

蛋白	8 個（室溫）
糖	半杯

做法

1. 將材料 A 的忌廉芝士及牛油打勻後，將蛋黃逐個加入打勻（蛋黃與忌廉芝士牛油一定要打至融合），繼而篩入餅粉。

2. 用另一個大盆，把材料 B 打勻後倒入（1）內，拌勻。

3. 再用另一個大盆，把蛋白打約十分鐘至企身，倒入（2）內輕輕將兩者混合。

4. 放入已預熱 160℃（325 ℉）焗爐內焗 1 小時即可。

POINT

蛋糕是否輕軟如棉花，蛋白佔非常重要的角色，但蛋黃與忌廉芝士牛油是否打發至融合亦是成功的關鍵，如一次倒入所有蛋黃，混合物會變成「豆腐渣」，焗後的蛋糕會很結實。

杏仁海綿蛋糕

ALMOND SPONGE CAKE
refer to p.164

材料 A		材料 C	
蛋黃	8 個	牛油溶液	半杯
糖	半杯		
		材料 D	
材料 B		蛋白	8 個
麵粉	1 杯		
發粉	1.5 茶匙		
杏仁粉	半杯		

做法

1. 快速打勻材料 A；轉慢速，將已篩勻的材料 B 逐少加入材料 A 內，拌勻。加入牛油溶液，拌勻。

2. 將蛋白打至企身，逐少加入以上的蛋糕料內，輕輕地捲勻。

3. 將蛋糕料倒入已抹油的焗盤內。

4. 放入已預熱 180℃（350 ℉）的焗爐內焗 25 分鐘即可。

CRISPY SESAME COOKIES

refer to p.165

香酥芝麻餅

材料 A

牛油溶液	半杯
麻油	2 湯匙
砂糖	3/4 杯
蛋	1 個
雲尼拿香油	半茶匙

材料 B

麵粉	1 杯
鹽	半茶匙
梳打粉	半茶匙
芝麻	3/4 杯

做法

1. 材料 A 拌勻後,加入蛋及雲尼拿香油,攪拌均勻。

2. 材料 B(除了芝麻)混合篩勻,逐少加入(1)內,最後加上芝麻拌勻。

3. 取出約 1 茶匙分量排在盤上,放入已預熱 170℃焗爐內焗 20 分鐘即可。

POINT　用白砂糖的芝麻餅味道較清甜;而用黃砂糖的味道會更香濃。

FRESH FRUIT SALAD

INGREDIENTS

1/2 papaya (peeled and sliced)

1 banana (peeled and sliced)

1 kiwi (peeled and sliced)

4 strawberries (sliced)

OTHER UTENSIL

1 wide-mouth glass jar (boiled in vigorously boiling water to sterilize)

DRESSING

1/2 cup plain yoghurt

2 tbsp honey

2 tbsp finely chopped mint leaves

2 tbsp grated orange zest

METHOD

1. In the glass jar, arrange the sliced papaya on the bottom. Spread some dressing on top. Then arrange a layer of sliced banana. Top with some dressing.

2. Arrange another layer of sliced kiwi. Spread some dressing over. Top with sliced strawberries. Serve.

POINT

1. Salad in a glass jar is all the rage lately and this is my first attempt. It turned out great! For a gathering of close friends at home, you can make individual portions in jars and each guest has his/her own jar.

2. I did the layering for presentation. But for a simpler variation, just toss the fruit and dress with the dressing.

3. You may use any fruit you like except those with high water content. Otherwise, the dressing will get watery.

SHREDDED SQUID COLD APPETIZER

INGREDIENTS

1 squid (Blanch in boiling water for 30 seconds. Turn off the heat and cover the lid. Leave it for 7 minutes. Soak in ice water. Cut into rings.)

1 sprig Chinese celery (cut into short lengths)

1/2 carrot (shredded)

40 g wood ear fungus (soaked in water until soft, shredded and blanched in boiling water)

DRESSING

1 tsp salt

1 tbsp sugar

1 tbsp ground yellow mustard

1 tbsp lemon juice

1 tsp sesame oil

METHOD

1. Mix all dressing ingredients.
2. Put all ingredients into a serving dish. Pour the dressing over and toss well. Serve.

POINT

You may use cuttlefish in place of squid for this recipe. Just make sure you shred the cuttlefish before use.

PORK LIVER COLD APPETIZER

INGREDIENTS

1 pork liver (about 400 g)

(Add 1 tsp Sichuan peppercorns and 1 chopped bird's eye chilli. Add enough water to cover. Leave it for 30 minutes. Drain.)

3 slices ginger

3 sprigs spring onion

DRESSING (MIXED WELL)

1 sprig coriander (use stems only, finely chopped)

1/2 tsp salt

2 tsp sugar

1 tbsp light soy sauce

1 tbsp aged vinegar

sesame oil

METHOD

1. Blanch the pork liver, ginger and spring onion in a pot of water. Bring to the boil. Turn off the heat and cover the lid. Leave them for 15 minutes. Drain. Wipe dry the pork liver with towel or paper towel. Slice it.

2. Drizzle the dressing over the pork liver. Serve.

MOUTH-WATERING CHICKEN

INGREDIENTS

1 chicken (Rub coarse salt weighing half as much as the chicken on the insides and outsides of the chicken. Then stuff the cavity with a clean towel. Wrap the chicken in a towel or paper towels. Refrigerate overnight.)

2 sheets fresh mung bean noodles (cut into thick ribbons)

4 thousand-year eggs (cut each into 1/8)

GARNISHES

1 cup chopped peanuts

2 tbsp finely chopped coriander

DRESSING (MIXED WELL)

3 tbsp sesame oil

3 tbsp aged vinegar

7 tbsp light soy sauce

3 tbsp brown sugar

1 tbsp finely diced ginger

1 tbsp grated garlic

1 tbsp finely chopped coriander

1 tbsp Sichuan pepper oil (adjust the amount according to your preferred piquancy)

METHOD

1. Remove the towel stuffed in the chicken. Rinse off all salt on it under running water. Steam over high heat for 25 minutes. Let cool and de-bone it. Shred the flesh with your hands.
2. Blanch the mung bean sheet noodles in boiling water until translucent. Drain.
3. Put the noodles into a serving plate. Top with shredded chicken. Arrange the thousand-year eggs along the rim of the plate.
4. Pour the dressing over. Sprinkle with garnishes on the dining table at last. Toss before eating.

POINT

1. Do not pour the steaming juices from the chicken down the drain. You can use it to cook rice instead of water. It gives the rice a meaty flavour and buttery texture.
2. Fresh mung bean sheet noodles could be hard to come by. You may use dried ones instead. Just soak them in water until soft. Blanch in boiling water until they turn translucent.

MRS TANG'S TARO CAKE

INGREDIENTS A

2 Cantonese dried pork sausages (about 85 g, steamed till done, diced)

I small piece Cantonese dried pork belly (about 85 g, steamed till done, diced)

4 dried shiitake mushrooms (soaked in water till soft; diced)

I tbsp dried scallop (soaked in water till soft, broken into shreds)

I tbsp Shaoxing wine

INGREDIENTS B

1/2 taro (about 900 g, peeled, grated coarsely into thick strips)

1/8 raw brown sugar slab

I tsp salt

I tsp five-spice powder

2 cups water

INGREDIENTS C

1/2 cup long-grain rice flour

1/2 cup glutinous rice flour

INGREDIENTS D

toasted sesames (use as much or as little as you like)

METHOD

1. Stir fry ingredients A in a wok with a little oil until fragrant. Set aside.

2. In the same wok, stir fry ingredients B until fragrant. Add ingredients A from step 1. Stir again.

3. Mix ingredients C and sieve together into the taro and dried meat mixture from step 2. Stir gently to mix well.

4. Pour the resulting batter into a greased tin. Steam over high heat for 30 minutes. Sprinkle with sesames. Serve.

POINT

1. The dried pork sausage and dried pork belly will be easier to dice after they are cooled completely after steamed.

2. After you rinse the sesames, just drain and put them into a cold wok. Turn the heat to medium and fry until dry and lightly browned.

DEEP-FRIED WONTONS WITH YAM BEAN AND SHRIMP FILLING

INGREDIENTS

1/2 yam bean (peeled and finely diced)

225 g shelled shrimps (Rub salt on them and rinse. Repeat once more. Wipe dry. finely chop them.)

1 pack instant shredded cod snack (about 30 g, finely chopped)

1 tbsp chopped coriander

150 g Cantonese wonton wrappers

1 egg yolk (whisked)

1 tbsp oyster sauce

METHOD

1. To make the filling, stir fry the shrimps over high heat in some oil for 10 seconds. Then fry the yam bean over high heat with the same oil. Put the shrimps back in. Add 1 tbsp of oyster sauce. Toss in shredded cod snack and coriander at last. Stir and set aside. Let cool completely.

2. Brush whisked egg yolk on the rims of each wonton wrapper. Put in some filling and fold the wonton wrapper into your preferred shape. Repeat until all ingredients are used.

3. Heat a pot of oil and deep fry the wontons over medium-low heat until golden. Serve.

POINT

1. Make sure you let the filling cool completely before wrapping it in wonton wrappers. Otherwise, the wonton wrappers may break easily.

2. As the filling was pre-cooked, just deep fry the wontons until the skin is crispy and golden.

3. After deep frying the wontons, leave them on paper towel to pick up excessive oil. They'd be less greasy that way.

DEEP-FRIED TARO BALLS

INGREDIENTS

1/4 taro (about 300 g, peeled and grated into fine shreds)
1/2 cup small peanuts (with skin on; roasted peanuts in an unoiled pan)
5 tbsp sugar
3 tbsp glutinous rice flour

METHOD

1. Add sugar to shredded taro and peanuts. Sprinkle with flour. Mix well and gently roll the mixture into small balls.
2. Deep fry in oil over low heat until cooked through. Serve.

POINT

Make sure you wear gloves when peeling the taro. Otherwise, your hands may get itchy after touching the taro.

DEEP-FRIED CUTTLEFISH CAKE WITH NORI SEAWEED

INGREDIENTS

12 slices Korean-style Nori seaweed

300 g seasoned minced cuttlefish (slapped on a chopping board repeatedly until sticky)

2 tbsp diced fatty pork

4 water chestnuts (peeled, crushed and finely chopped)

1 sprig coriander (with the stems chopped, and leaves set aside)

1/2 cup sesames

METHOD

1. Add water chestnuts, diced fatty pork and coriander stems to the minced cuttlefish. Stir well and slap it on a chopping board hard for a few times. Add sesames and stir well. Smear a thin layer of minced cuttlefish mixture on each sheet of Nori seaweed. Put a coriander leaf on cuttlefish side.

2. Heat oil up to 40°C. Slide in cuttlefish cake one by one with the cuttlefish side facing down. Swirl it continuously to heat through evenly. Flip it after the cuttlefish is cooked through. Turn to high heat and fry for 20 more seconds. Drain and serve.

POINT

You may put in a piece of cuttlefish cake to test to temperature. It's hot enough as long as you see small bubbles coming out around the cuttlefish cake. If the oil is too hot, the Nori seaweed tends to burn very quickly.

STEAMED RADISH DUMPLINGS

INGREDIENTS

1 radish (about 1.2 kg, peeled, grated into shreds, marinated with 1 tsp of salt)
1 1/2 cups long-grain rice flour
3 tbsp glutinous rice flour
1 tsp ground white pepper
2 tbsp Shaoxing wine
1/2 cup sesames (as garnish)

ASSORTED PRESERVED MEAT

1 Cantonese dried pork sausage (steamed till soft, finely chopped)
1/4 Cantonese dried pork belly (steamed till soft, finely chopped)
1 tbsp dried shrimps (soaked in water till soft, finely chopped)
2 dried shiitake mushrooms (soaked in water till soft, finely chopped)
1 clove garlic

METHOD

1. Rinse the sesames with water. Drain well. Toast in a dry pan until golden and fragrant.
2. Decant any liquid drawn out of the shredded radish. Sieve in long-grain rice flour, glutinous rice flour and ground pepper. Stir well.
3. Heat wok until hot. Add cold oil. Stir fry the assorted preserved meat until fragrant. Transfer into the flour-radish mixture from step 2. Mix well. Shape into balls and transfer into a steamer. Steam over high heat for 12 minutes. Turn off the heat and keep the lid on. Leave the dumplings in the steamer for 3 more minutes. Sprinkle with sesames on top. Serve.

POINT

1. It's hard to grate a radish as it keeps rolling around. If you cut it in half, that would make it easier.
2. I marinated the shredded radish with salt because the salt draws out juices and make it less bitter.
3. When you stir fry the assorted preserved meat ingredients, heat the wok and add cold oil. Then put in the ingredients right away so that they have time to warm up alongside the oil slowly. They are less likely to burn that way.

127

DUMPLINGS WITH CHAYOTE PORK FILLING

INGREDIENTS

2 chayotes (about 550 g, peeled, shredded, marinated with 1 tsp of salt, with any liquid decanted)

250 g ground pork

2 dried shiitake mushrooms (soaked in water till soft, finely chopped)

300 g round dumpling skin

1 egg yolk (for sealing the dumpling skin)

SEASONING

1/2 tsp salt

1/2 tsp sugar

1 tsp caltrop starch

1 tsp sesame oil

1 egg white

METHOD

1. To make the filling, add seasoning to the pork. Stir well and slap the pork on a chopping board hard for a few times. Add shredded chayote and shiitake mushrooms. Mix well.

2. Lay flat a piece of dumpling skin. Put a teaspoon of filling on it. Brush some whisked egg yolk on the rim of the dumpling skin. Pinch the rim together into dumplings.

3. Boil a pot of water and put in the dumplings. Cook until they float. Add 1/2 cup of cold water and bring to the boil again. Drain and save the dumplings in serving dish. Serve with vinegar or light soy sauce on the side.

POINT

1. Add salt to the shredded chayote to draw the moisture out of it. But before you add it to the filling, just decant any liquid drawn out of it, without squeezing it completely dry. Otherwise, the filling will be too stiff and not juicy enough.

2. To make sure the dumpling filling is properly cooked through, I always add 1/2 cup of cold water to the dumplings after they float. When the water boils again, the dumplings are perfectly cooked without being mushy.

TEA-SCENTED CHICKEN WINGS

INGREDIENTS

900 g mid-joint chicken wings

3 cups brine (1/2 cup of coarse salt dissolved in 3 cups of boiling water)

8 slices ginger

2 tbsp osmanthus sauce (method follows)

1/2 cup osmanthus tea (black tea leaves and dried osmanthus flowers steeped in hot water; strained)

OSMANTHUS SAUCE

8 tbsp Hoi Sin Sauce

4 tbsp Chu Hau Sauce

2 tbsp candied osmanthus

1 tbsp ground bean sauce

1 tbsp creamy peanut butter

3 cubes fermented tofu (mashed)

1/2 raw brown sugar slab (crushed)

METHOD

1. To make the osmanthus sauce, put all ingredients into a pot. Cook over medium-low heat while stirring continuously. Bring to the boil and turn off the heat. Let cool and transfer into an airtight glass jar sterilized with vigorously boiling water.

2. Boil the brine and put in the chicken wings. Bring to the boil over high heat and turn off the heat. Cover the lid and leave them for 10 minutes. Rinse the chicken wings in running cold water for 30 minutes. Drain.

3. Heat a wok and add 1 tbsp of oil. Stir fry sliced ginger over high heat until fragrant. Put in the chicken wings and osmanthus sauce. Toss well. Add the osmanthus tea. Cook over high heat until it almost dries out. Serve.

POINT

1. You can add chicken liver to this recipe. Just make a cut on each lobe so that it cooks more quickly and picks up the seasoning better.

2. For the detailed steps to prepare frozen chicken wings, please refer to p.22 of another title under the same author (Gigi's Kitchen).

BRAISED DUCK IN BEER SAUCE

INGREDIENTS

1 duck (rinsed, cut into chunks, blanched in boiling water)
8 red dates

AROMATICS

2 cloves star-anise
2 pieces ginger
4 cloves garlic
1 bird's eye chilli
2 cubes rock sugar

SEASONING

2 cans beer
1 tsp salt
1 tbsp dark soy sauce

METHOD

1. Heat a wok till hot and add cold oil. Stir fry the aromatics. Put in the duck. Fry over high heat until fragrant.
2. Add seasoning and red dates. Simmer for 2 hours. Serve.

POINT

You can shorten the cooking time to 30 minutes if you use a pressure cooker.

BRAISED CHICKEN WITH CHESTNUTS AND GARLIC

INGREDIENTS

4 boneless chicken thighs (cut into pieces, mixed with marinade)

12 single-clove garlic (peeled)

18 shelled chestnuts (blanched in boiling water for 10 minutes, set aside the blanching water for braising chicken)

AROMATICS

2 slices ginger

1 red chilli

1 whole pod star anise

2 cubes rock sugar

MARINADE FOR CHICKEN

1/2 tsp salt

1 tsp sugar

1 tbsp light soy sauce

1 tbsp Shaoxing wine

1 tsp sesame oil

ground white pepper

SEASONING

2 tbsp light soy sauce

2 tbsp oyster sauce

1 tbsp sugar

METHOD

1. Stir fry the aromatics in a wok with some oil until fragrant. Put in the chicken and fry over high heat till oil is rendered out of the skin. Put in the garlic and fry well.
2. Put in the chestnuts and add seasoning. Add some water from blanching chestnuts if necessary. Cover the lid and cook for 3 minutes. Serve.

POINT

1. Shelling and peeling chestnuts are such a tedious task. Some people say you can cut off the hilum of chestnuts and heat them up in a microwave oven for 2 to 3 minutes. That would make them a lot easier to shell and peel.
2. For variation, you can use duck instead of chicken by following the same step. Just make sure you add enough water to cover the duck and simmer for 45 minutes before serving.

ROLLING BRAISED CHICKEN

INGREDIENTS

1 chicken (rinsed, wiped dry, with 1 tsp of salt rubbed on the insides)

SEASONING

1/2 cup Shaoxing wine

1/3 cup Maggi's seasoning

2/3 raw brown sugar slab

4 slices ginger

2 sprigs spring onion

1/2 cup boiling water

METHOD

1. Heat a wok and add 2 tbsp of oil. Put in the chicken and keep rolling it in the wok to brown the skin evenly.

2. Add the seasoning and turn to medium heat. Roll the chicken to have the breast facing down. Cover the lid and cook for 5 minutes. Open the lid and keep rolling the chicken for about 20 minutes until the juices reduce and turn thick. Serve.

POINT

This rolling method makes sure the chicken picks up the colour and seasoning evenly.

ROMAN-STYLE CHICKEN STEW

INGREDIENTS

6 chicken fillets (marinated with 1 tsp salt and 1 tsp of ground black pepper)

1 large onion (finely shredded)

4 tomatoes (cut into wedges)

1/2 red bell pepper

1/2 yellow bell pepper ⎫ — (with stems cut off, de-seeded, cut into chunks)

1/2 green bell pepper

1 tbsp capers (crushed)

3/4 cup white wine

1 tsp sea salt

2 tsp crushed black pepper

4 tbsp olive oil

METHOD

1. Add oil to a cold pan. Put in the shredded onion. Turn on to medium heat and sweat the onion until transparent. Put in the chicken. Fry over high heat for 10 minutes. Pour in the white wine and cook for 5 minutes. Remove the chicken and set aside.

2. When the onion is mushy, add tomatoes, bell peppers and capers. Stir well. Season with salt and ground black pepper. Cover the lid and cook for 20 minutes until all ingredients are tender. Put the chicken back in and heat it through. Serve.

POINT

This is a rustic home-style dish I learned from the moms in the Italian countryside last summer. It may not look too fancy, but it tastes awesome.

STEAMED PUMPKIN STUFFED WITH CHICKEN AND ONION

INGREDIENTS

1 Japanese pumpkin (about 2 kg, with the top quarter cut off, de-seeded)

3 boneless chicken thigh (cut into pieces, seasoned with some salt and ground white pepper)

1/2 onion (finely chopped)

2 tbsp osmanthus sauce

1 sprig spring onion (finely shredded)

METHOD

1. Rub osmanthus sauce on the insides of the pumpkin and the top.
2. In a wok, stir fry onion in some oil until fragrant. Add chicken and the remaining osmanthus sauce. Cook until the chicken is done. Transfer the mixture into hollowed-out pumpkin.
3. Put the pumpkin into a big soup bowl. Steam for 20 minutes.
4. Sprinkle with finely shredded spring onion. Sizzle with hot cooked oil. Serve.

POINT

1. For the method of osmanthus sauce, please refer to the recipe of Tea-scented chicken wings on P.129.
2. The big bowl helps hold the pumpkin's shape. The pumpkin also tends to be more tender that way.

CRAB CAKE

INGREDIENTS

1 crab (steamed for 10 minutes; shelled and de-boned)

1/2 onion (diced)

2 shallots (finely chopped)

4 slices smoked bacon (finely chopped)

6 button mushrooms (finely chopped)

1/2 cup peas

4 water chestnuts (crushed)

2 tbsp creamy salad dressing

1 tbsp grated parmesan

3 eggs (whisked)

METHOD

1. Heat a wok and add oil. Stir fry onion and shallot over medium heat until fragrant. Add bacon, button mushrooms, peas, water chestnuts and crab meat. Stir well. Set aside.

2. Add salad dressing and grated parmesan. Stir well. Stir in the eggs at last.

3. Scoop out one tablespoon of the crab cake mixture. Fry both sides over low heat in a little oil until browned. Serve.

POINT

1. If you decide to serve crabs in whole, keep them in the freezer to sedate them before steaming. Otherwise, live crabs may struggle when subject to heat and lose their legs.

2. Clean the crab very well with a toothbrush before cooking or steaming it.

FRIED PRAWNS IN OSMANTHUS AND DISTILLERS GRAIN SAUCE

INGREDIENTS

900 g medium prawns (shelled, deveined, cut along the back, rub salt on them twice and rinse after each rubbing, wipe dry)

1/2 yam bean (peeled, cut into thick strips, arrange on a serving plate)

AROMATICS

2 dried shiitake mushrooms (soaked in water till soft; finely chopped)

4 slices ginger (finely chopped)

4 cloves garlic (grated)

2 sprigs spring onion (finely chopped)

SAUCE

2 tbsp osmanthus sauce

2 tbsp distillers grains

1 tbsp light soy sauce

1 tsp aged vinegar

1 tbsp sugar

sesame oil

METHOD

1. Heat a little more oil than you would for pan frying. Shallow fry the prawns until just done. Drain and set aside.

2. With the remaining oil in the wok. Stir fry the aromatics over medium heat until fragrant. Sizzle with wine. Put in the sauce ingredients and bring to the boil. Put the prawns back in. Toss to coat the prawns evenly in the sauce. Transfer on the serving plate over the bed of yam bean. Serve. Raw yam bean helps alleviate Dryness-Heat from Chinese medical point of view.

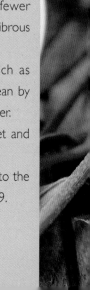

POINT

1. Pick yam beans with fewer stripes and fewer roots. They tend to be crunchier and less fibrous that way.

2. As opposed to peeling other tubers such as taro or sweet potato, you can peel yam bean by tearing it off with your hands, and not a peeler.

3. Yam bean can be eaten raw. It tastes sweet and crunchy.

4. To make the osmanthus sauce, please refer to the recipe of Tea-scented chicken wings on P.129.

STIR-FRIED CLAMS WITH CHILLIES AND SHISO

INGREDIENTS

600 g live clam (blanched in boiling water until the clams open)

AROMATICS

2 slices ginger (finely diced)

4 cloves garlic (grated)

I red chilli (finely chopped)

2 tbsp fermented black beans (rinsed)

4 fresh shiso leaves (a.k.a. perilla leaves, finely chopped)

SEASONING

I tbsp light soy sauce

I tbsp oyster sauce

I tsp sugar

water

METHOD

1. Heat a wok and add oil. Stir fry the aromatics in the order listed. Put in the clams and stir well.
2. Add seasoning and stir again. Serve.

POINT

1. All shellfish should be blanched in boiling water until they open before they are used in stir-fries, just to make sure there isn't any stale or dead one in the batch. The shellfish also tends to be more succulent and less likely to overcook this way, as you're simply coating it in the sauce in the stir-frying stage.
2. Shiso leaves detoxify the body and remove the fishy taste of the ingredients. They are commonly used in shellfish and crab dishes.

WINE-SCENTED TURBOT FILLET IN CLAY POT

INGREDIENTS

1 turbot (about 900 g, rinsed and scaled)

4 tbsp Shaoxing wine

2 red chillies (sliced diagonally)

AROMATICS

5 slices old ginger

8 shallots

8 cloves garlic

5 sprigs spring onion (cut into short lengths about 2 inches long)

MARINADE

1 tsp salt

ground white pepper

1 tsp ginger juice

spring onion juice (crush the white part of spring onion the back of a knife and squeeze out the juices)

1 tsp caltrop starch

1 tbsp light soy sauce

METHOD

1. De-bone the fish and slice the fillet thickly. Add marinade and mix well.

2. Stir fry the aromatics over medium heat in a little oil until fragrant. Transfer into a clay pot. Arrange the fish fillet on top. Cover the lid and pour 4 tbsp of Shaoxing wine along the rim of the lid. Turn to low heat and cook for 4 minutes until the fish is done. Sprinkle with red chillies. Serve in the clay pot.

POINT

1. When you de-bone the fish, make sure you slide the knife as closely to the bones as you can.

2. You may also marinate the fish bones and pan fry them. It makes a great snack to go with alcohol.

STEAMED SHIITAKE MUSHROOM STUFFED WITH MINCED SHRIMP FILLING

INGREDIENTS

12 dried shiitake mushrooms (pick those about the same sizes; prepared as outlined in tips)

250 g shrimps (deveined, rinsed and rubbed in water twice; wiped dry and crushed)

1 small piece fatty pork (finely chopped; about 1 tbsp)

3 water chestnuts (peeled, crushed and finely chopped)

200 g leafy greens (rinsed)

1 tbsp grated Jinhua ham (as garnish)

1 tbsp oyster sauce

MARINADE

1 tbsp egg white

ground white pepper

sesame oil

METHOD

1. Add marinade to the minced shrimps. Lift the minced shrimp mixture and slap it hard on a chopping board for a few times. Add water chestnuts and fatty pork. Mix well. Wipe dry the shiitake mushrooms. Dust the gills lightly with caltrop starch. Stuff each shiitake mushrooms with the filling. Sprinkle with grated Jinhua ham. Steam over high heat for 5 minutes.

2. Boil some water. Add 1 tbsp of oil and 1 tsp of salt. Blanch the leafy greens in water for 2 minutes. Drain. Put the leafy greens on the side of the serving plate. Drizzle with oyster sauce. Serve.

POINT

1. To prepare dried shiitake mushrooms, soak them in water until soft. Transfer into a microwave-safe bowl and strain the soaking water. Pour the strained soaking water back in with the mushrooms and add water to cover if needed. Add 2 slices of ginger, 2 sprigs of spring onion and 1/4 raw brown sugar slab. Cover with microwave cling film and pierce two holes on it. Heat in a microwave oven over high power for 10 minutes. Then heat again for 10 more minutes. Heat some oil in a wok and stir fry 2 cloves of garlic until fragrant. Put in the mushrooms and stir well. Add 2 tbsp of straw mushroom-flavoured dark soy sauce and the soaking water. Cook until the liquid reduces. Add 2 tbsp of oyster sauce and stir well. Turn off the heat and let cool. Divide the shiitake mushrooms among storage boxes or zipper bags. Keep in the freezer. They don't need to be thaw before use. Just put them straight into a pot to reheat and serve.

2. Make sure you wipe the shiitake mushrooms completely dry. Otherwise, it would be hard for the caltrop starch to adhere to them.

FRIED PRAWNS WITH CHILLI BLACK BEAN SAUCE

INGREDIENTS

600 g medium prawns (antennae trimmed; deveined)

4 red chillies (rinsed, sliced diagonally)

AROMATICS

2 tbsp fermented black beans

1 tbsp grated garlic

1 tbsp finely diced ginger

1 tsp sugar

1 tbsp oil

SEASONING

1 tbsp light soy sauce

1 tbsp oyster sauce

2 tsp sugar

METHOD

1. Wipe dry the prawns with towel or paper towel. Blanch in oil briefly. Drain.
2. Mix all aromatics.
3. Heat a wok and add oil. Stir fry aromatics over medium heat until fragrant. Put in the chillies and toss well. Put in the prawns and stir again. Add seasoning and stir to coat all prawns in the sauce evenly. Serve.

POINT

Make sure you wipe the prawns completely dry before blanching them in oil. Otherwise, the oil may splatter and it's very dangerous.

POACHED SLICED GRASS CARP IN SOUR SPICY BROTH

INGREDIENTS

1 grass carp fillet (sliced, marinated for 30 minutes)

4 slices ginger (finely shredded)

1/2 cup shredded bamboo shoot pickle

1 red chilli (finely chopped)

1 tbsp chilli sauce (use as much or as little as you like)

1 1/2 cup boiling water

1 tsp sugar

1 sprig coriander (finely chopped)

MARINADE FOR GRASS CARP

1/2 tsp salt

1 tbsp caltrop starch

1/3 tsp sugar

1 egg white

ground white pepper

METHOD

1. Heat a wok and add 2 tbsp of oil. Stir fry ginger, bamboo shoot pickle, red chilli and chilli sauce until fragrant. Add boiling water to cover all ingredients.

2. Add 1 tsp of sugar. Put in the marinated grass carp and spread them apart. Bring to the boil and turn off the heat. Sprinkle with coriander on top. Serve.

POINT

Marinating the fish with egg white helps keep it juicy and tender.

|42 SPINACH FISH ROLLS

INGREDIENTS

1 mandarin fish (de-boned and sliced; marinated with salt and ground white pepper)

600 g spinach (rinsed; separate the leaves and stems)

1 tbsp Shaoxing wine

3 sprigs spring onion (blanched in boiling water)

3 cups water

SAUCE

1/2 cup milk

2 tbsp pine nuts (toasted in a dry wok)

1/2 tsp salt

ground white pepper

METHOD

1. Fry the fish bones, tail and head in some oil with a slice of ginger until fragrant. Add boiling water and cook for 10 minutes for the flavours to infuse. Strain and set aside the stock.

2. Cut the spinach stems into short lengths. Roll a slice of fish around some spinach stems. Tie a sprig of spring onion firmly around to secure. Cook the fish rolls in the stock from step 1 for 2 minutes. Save the fish rolls on a serving plate.

3. Blanch the spinach leaves very briefly in the stock. Transfer the stock and spinach leaves into a blender. Add the sauce ingredients and blend until fine. Drizzle over the fish rolls. Serve.

POINT

1. For this recipe, pick a mandarin fish that is big, with firm and thick flesh.

2. The rules of thumb for pan-frying fish: add cold oil to a red hot wok, and patience matters. Fry the fish on one side until golden before flipping it. If you keep flipping it repeatedly, the flesh will fall off and your fish looks hideous.

3. You may use blanched chives in place of spring onion to tie the fish rolls.

PAN-FRIED PORK PATTIES WITH LOTUS ROOT AND DRIED TANGERINE PEEL

INGREDIENTS

1 small segment lotus root (about 300 g, with skin scraped off; diced very finely)

300 g ground pork (marinated)

1 piece dried tangerine peel (soaked in water till soft; finely diced)

MARINADE FOR PORK

1/2 egg

1 tsp salt

1/2 tsp sugar

1 tsp Shaoxing wine

ground white pepper

grated ginger

finely chopped coriander

finely chopped spring onion

METHOD

1. Blanch the diced lotus root in boiling water briefly. Drain. Rinse in cold water. Drain and set aside.
2. Stir diced lotus root and dried tangerine peel into the marinated ground pork. Grease your hands and squeeze the ground pork mixture into balls. Then flatten them into round patties.
3. Fry the patties in a pan over medium heat until both sides golden. Serve.

POINT

This is a great dish to go with rice, but you can also serve it as a snack. Just wrap it in pancake and serve it with Worcestershire sauce as a dip. Simply divine.

STIR-FRIED SLICED LAMB WITH FRESH YAM

INGREDIENTS

450 g sliced lamb

300 g fresh yam (sliced, blanched in boiling water for 20 seconds to remove the slime)

1 sprig Chinese celery (with leaves removed, cut into short lengths)

4 slices ginger

4 cloves garlic (gently crushed)

2 sprigs spring onion (cut into short lengths)

2 red chillies (finely chopped)

SEASONING

1 tsp salt

1 tbsp light soy sauce

2 tbsp wine

ground white pepper

sesame oil

METHOD

1. Heat a wok till hot and add a little cold oil. Sear the lamb over high heat until juices come out. Drain the liquid. Set aside.

2. In the wok, stir fry ginger, spring onion and red chillies until fragrant. Add the yam, Chinese celery and lamb. Add seasoning and toss well. Serve.

POINT

1. When you peel the yam, it's advisable to wear gloves because its juices may cause itchiness when in touch with skin.

2. After you peel the yam, you should keep it in salted water. It tends to oxidized and blacken easily when exposed to air.

BITTER MELON STUFFED WITH GROUND PORK FILLING

INGREDIENTS

1 long bitter melon (cut into rings about 1 inch thick, de-seeded)

120 g ground pork (marinated)

1 tbsp fermented black beans (rinsed with water, mixed with 1 tsp grated garlic, 1 tsp finely chopped soaked dried tangerine peel, and 1 tsp sugar)

MARINADE FOR PORK

1 tbsp ginger wine (Squeeze the juice out of 1 tbsp of grated ginger and mix it with 1 tbsp of Shaoxing wine)

1 tsp light soy sauce

1 tsp caltrop sauce

1 tsp sesame oil

ground white pepper

SAUCE

1/2 tsp salt

1/2 tsp sugar

1 tbsp oyster sauce

2 tbsp water

METHOD

1. Add water to a pot until about 2 inches deep. Bring to the boil. Add 1 tsp of oil and 1 tsp of salt. Put in the bitter melon and cook for 2 minutes. Drain and soak in cold water for 1 minute. Drain again.

2. Mix the pork with marinade. Dust the insides of the bitter melon rings with caltrop starch. Stuff each bitter melon ring with the pork filling. Fry the stuffed bitter melon in oil until both sides golden for about 2 minutes.

3. In another wok, heat 2 tbsp of oil and stir fry fermented black beans until fragrant. Put in all sauce ingredients and bring to the boil. Pour the sauce over the bitter melon in the other wok. Toss well and serve.

SPICY BRAISED PORK TROTTER

INGREDIENTS

1 pork trotter (chopped into pieces; blanched in boiling water with the spice mix; rinsed in cold water)

SPICE MIX FOR BLANCHING PORK TROTTER

1 piece cassia bark
2 pods star anise
3 bay leaves
3 slices liquorice
*put into a muslin bag and tie well

AROMATICS

2 slices ginger
4 cloves garlic
1 tbsp Sichuan peppercorns
1 bird's eye chilli

SEASONING

1 tbsp Shaoxing wine
1/2 raw brown sugar slab
1 tbsp dark soy sauce
1/2 tsp salt
1 cup boiling water

METHOD

1. Heat a wok and add oil. Stir fry the aromatics until fragrant. Put in the blanched pork trotter. Fry until all sides golden. Add seasoning.
2. Cover the lid and simmer for 45 minutes. Or, cook for 15 minutes in a pressure cooker.

POINT

I put the spices into a muslin bag, so that you don't have to fish out the spices afterwards. Your guests won't bite into any spice by accident either.

DEEP-FRIED CHITTERLINGS WITH SWEET AND SOUR SAUCE

INGREDIENTS

2 frozen pork large intestines (cooked in water with 4 slices ginger, 2 sprigs spring onion for 1 hour; sliced; with fat trimmed off)

DEEP FRYING BATTER

8 tbsp caltrop starch

1 tsp flour

2 egg whites

1/4 cup cold water

*Slowly stir cold water into the rest of the ingredients. Stir to mix well.

AROMATICS

2 slices ginger

2 sprigs spring onion

2 cloves garlic

SWEET AND SOUR SAUCE

1/2 tsp salt

1 tbsp soy sauce for steamed fish

1/2 tsp caltrop starch

1 tbsp sugar

1 tbsp Zhenjiang black vinegar

2 tbsp water

METHOD

1. Put pork intestines into the deep frying batter. Mix well.

2. Heat a pot of oil. Deep fry the pork intestines once over low heat. Set aside.

3. Heat the pot of oil over high heat. Deep fry the pork intestines once more.

4. In a wok, stir fry ginger, spring onion and garlic until fragrant. Put in all sauce ingredients. Bring to the boil. Toss the pork intestine to coat evenly. Serve.

POINT

You may use fresh pork intestines instead of frozen ones though that's more work involved to prepare them. Rub salt and vinegar on the fresh pork intestines for a few times. Turn the intestines inside out. Trim off the fat. Rub salt and vinegar on it again for a few more times. Turn the intestines back in. Rub again and rinse well. Boil a pot of water and blanch the intestines for 1 minute. Drain. Simmer the intestines with ginger and spring onion for 1 hour. Remove and cut into short lengths. Set aside.

BRAISED BOK CHOY WITH DRIED SHRIMPS

INGREDIENTS

600 g Bok Choy (rinsed)
1 tbsp finely diced ginger
1 tbsp dried shrimps (rinsed)

METHOD

1. Heat a wok and add oil. Stir fry ginger and dried shrimps over high heat till fragrant. Add 1 cup of boiling water. Cook for 3 minutes for flavours to infuse.

2. Put in the Bok Choy and cook for 2 minutes. Serve.

POINT

1. You don't need to soak the dried shrimps till soft for this recipe. Just rinse it and drain. Otherwise, they may not be flavourful enough.

2. If you happen to make Mouth-watering Chicken (recipe on P.122), you may also cook Bok Choy with the steaming juices from the chicken. It's sinfully delicious.

VEGETARIAN TOFU PANCAKES

INGREDIENTS

1/2 onion (finely chopped)

1 tbsp grated garlic

4 water chestnuts (peeled and crushed)

2 dried shiitake mushrooms (soaked till soft, finely chopped)

1 whole head fresh lily bulb (finely chopped)

1 tbsp walnuts ⎤

1 tbsp peanuts ⎬— ground together

2 tbsp sesames ⎦

2 cubes soft tofu (mashed)

SEASONING

1 tbsp caltrop starch

1/2 tsp salt

1 tsp ground white pepper

METHOD

1. Stir fry onion and garlic in oil until fragrant. Put in water chestnuts, shiitake mushrooms and lily bulb in that particular order. Stir well.

2. Add assorted ground nuts. Stir again. Put in mashed tofu and seasoning. Mix well. Shape into small round patties.

3. Heat a little oil in a pan. Fry the patties over low heat until both sides golden.

POINT

You may serve the pancakes with Thai sweet chilli sauce.

STIR-FRIED MUNG BEAN VERMICELLI WITH WOOD EAR AND DRIED TOFU

INGREDIENTS

2 bundles mung bean vermicelli

SEASONING

1 tsp salt

1 tbsp oil

ground white pepper

1 tbsp dark soy sauce

ASSORTED VEGETABLES

2 shallots (shredded)

15 g wood ear fungus (soaked in water till soft, finely shredded)

30 g salted radish (finely shredded)

1 slice five-spice dried tofu (finely shredded)

1/2 carrot (coarsely grated)

50 g snow peas (finely shredded)

1 sprig Chinese celery (shredded)

1 red chilli (shredded)

METHOD

1. Boil a pot of water and put in the seasoning. Cook the mung bean vermicelli until soft. Drain. Cut the vermicelli into half their lengths.

2. In a wok, stir fry the assorted vegetables in some oil in the order listed. Add 1/2 tsp of salt and put in the mung bean vermicelli. Toss quickly to mix all ingredients well. Serve. If you find the noodles too dry in the process of stir frying, add some stock at a time.

POINT

1. Blanching the mung bean vermicelli in seasoning instead of soaking them in hot water gives them some background seasoning so that they are more flavourful. The vermicelli also tend to be chewier in texture and less likely to break down into bits.

2. Well-made stir-fried vermicelli should not be too wet or too dry. If you find the noodles too dry, add some stock.

BLACK SILKIE CHICKEN SOUP WITH SHI HU

INGREDIENTS

1 black silkie chicken (dressed, rinsed and blanched in boiling water)

20 g Shi Hu

20 g American ginseng

40 g Hai Zhu Tou (or Yu Zhu)

40 g dried Tu Fu Ling (or 160 g fresh Tu Fu Ling)

4 dried figs

80 g dried conches

120 g cashew nuts

5 slices ginger

20 cups water

METHOD

1. Rinse the dried conches. Soak them in water overnight. Transfer into a rice cooker. Add 1 cube of rock sugar. Turn on the cooker and cook until the liquid dries out.

2. Put all ingredients into a pot. Bring to the boil over high heat and keep boiling for 10 minutes. Turn to medium-low heat and simmer for 3 hours.

3. Season with salt to taste. Serve.

POINT

1. After you rehydrate the dried conches, let cool and divide among zipper bags. Keep in the freezer and thaw before use.

2. From Chinese medical point of view, Hai Zhu Tou clears Heat, nourishes the Yin, promotes body fluid secretion, and moistens Dryness and the Lungs.

3. This soup nourishes the Yin, clears Heat, promotes body fluid secretion and benefits the Stomach. It's recommended to those who have to work late hours, with dry mouth, poor sleep quality, diabetes or high blood pressure.

CRUCIAN CARP SOUP WITH PUMPKIN

INGREDIENTS

1 crucian carp (rinsed and fried in oil until browned)

1/2 pumpkin (about 900 g, peeled, seeded and cut into chunks)

300 g pork shoulder blade (blanched in boiling water, drained)

2 dried figs

2 candied dates

2 slices ginger

18 cups water

OTHER UTENSILS

muslin bag

METHOD

1. Put the fried fish into the muslin bag and tie well.

2. Put all ingredients into the pot. Bring to the boil over high heat and keep boiling for 10 minutes. Turn to low heat and simmer for 3 hours.

3. Season with salt to taste. Serve.

POINT

1. When you make fish soup, it's advisable to put the fish in a muslin bag before cooking it with other ingredients. You can rest assured that there won't be fish bone in the soup that way.

2. No matter you fry the fish in oil or blanch it in boiling water, adding a couple slices of ginger helps make the fish less fishy.

3. From Chinese medical point of view, crucian crap expels Dampness, nourishes the Yin, promote Qi (vital energy) and blood circulation.

BEEF SOUP WITH BUTTON MUSHROOMS

INGREDIENTS

450 g beef (coarsely diced)

1 tbsp butter

6 cups water

VEGETABLES

225 g button mushrooms (coarsely diced)

1 onion (finely diced)

1/2 carrot (peeled and finely diced)

2 celery stems (finely diced)

1 potato (peeled and finely diced)

SEASONING

1/4 tsp dried thyme

1/2 cup white wine

2 tbsp tomato paste

METHOD

1. Sear beef in butter. Add seasoning and stir well. Set aside.

2. In another wok, stir fry all vegetables (except potato) for 3 minutes. Add the beef and seasoning mixture from step 1. Add water and boil for 50 minutes.

3. Put in the diced potato at last and boil for 10 minutes. Taste the soup and season with salt accordingly. Serve.

FISH HEAD SOUP WITH DANG SHEN

INGREDIENTS

1 head of bighead carp (cut into halves)

300 g pork bones (blanched in boiling water)

40 g Dang Shen

40 g Bei Qi

30 g Tian Ma

10 water chestnuts (peeled)

10 red dates (pitted)

2 slices ginger

14 cups boiling water

4 tbsp rice wine

METHOD

1. Fry the fish head in oil until both sides golden. Sizzle with 2 tbsp of rice wine. Add 2 cups of boiling water.

2. Transfer the mixture from step 1 into a big pot. Add the remaining water and ingredients. Boil over low heat for 2 hours. Turn off the heat. Add the remaining 2 tbsp of rice wine.

3. Season with salt to taste. Serve.

POINT

From Chinese medical point of view, this soup benefits the brain, regulates blood chemistry and expels Wind.

PORK SOUP WITH BAMBOO SHOOTS

INGREDIENTS

2 fresh bamboo shoots (peeled, cut into chunks)

600 g pork shoulder blade (blanched in boiling water)

300 g dried soybeans (rinsed, soaked in water)

1 pickled mustard green (rinsed, sliced)

150 g dried cuttlefish (rinsed, skin peeled off)

1 whole dried tangerine peel (soaked in water till soft)

3 candied dates

5 sliced ginger

18 cups water

METHOD

1. Put all ingredients into a pot. Bring to the boil over high heat and keep boiling for 5 minutes. Turn to medium-low heat and simmer for 2 hours.
2. Season with salt to taste. Serve.

POINT

It's not difficult to prepare fresh bamboo shoots. Just make a light incision along the length. Then peel off the leaves one by one until you see the edible core. Cut off the base of the bamboo shoots which could be too fibrous to eat.

OXTAIL SOUP WITH DANG SHEN AND BEI QI

INGREDIENTS

1 oxtail (blanched in boiling water)

20 g Dang Shen

20 g Bei Qi

20 g Ba Ji

20 g Du Zhong

30 g Tu Fu Ling

30 g processed fox nuts

3 Tsaoko fruits

3 bay leaves

1 whole dried tangerine peel (soaked in water till soft)

5 candied dates

5 slices ginger

20 cups water

METHOD

1. Put all ingredients into a pot. Bring to the boil over high heat and keep boiling for 5 more minutes. Turn to medium-low heat and simmer for 3 hours.

2. Season with salt to taste.

POINT

From Chinese medical point of view, this soup helps strengthen the Kidneys and secure Jing (the essence of life).

LONGLI LEAF AND CHUAN BEI MU SOUP

INGREDIENTS

600 g lean pork (blanched in boiling water)

30 g longli leaves

30 g Chuan Bei Mu

30 g sweet and bitter almonds

30 g dried lily bulbs

30 g Hai Zhu Tou (or Yu Zhu)

30 g lotus seeds

30 g dried figs

60 g dried oysters

1 whole dried tangerine peel (soaked in water till soft, pith scraped off)

5 slices ginger

20 cups water

METHOD

1. Put all ingredients into a pot. Bring to the boil over high heat. Keep boiling for 5 more minutes. Turn to medium-low heat and simmer for 3 hours.
2. Season with salt to taste.

POINT

1. You can get longli leaves from herbal stalls in wet markets.
2. Chuan Bei Mu clears Heat, dissipates phlegm, moistens the Lungs and stops cough.
3. From Chinese medical point of view, this soup soothes the Dryness in the Lungs, quenches thirst and dissipates phlegm.

158 CREAM OF PUMPKIN SOUP WITH ASSORTED MUSHROOMS AND TOFU

INGREDIENTS

1/2 pumpkin (peeled and sliced)

4 fresh shiitake mushrooms ⎤
6 button mushrooms ⎥ wiped clean with damp cloth, finely
1 pack Shimeji mushrooms ⎥ chopped and blanched in boiling water
1 pack tea tree mushrooms ⎦

1 cube soft tofu (finely chopped)

1 tbsp evaporated milk

4 cups water

1 tsp salt

METHOD

1. Steam pumpkin over high heat for 8 minutes. Transfer into a blender. Add 1/2 cup of water. Blend until fine.

2. Boil 4 cups of water in a pot. Put in the pumpkin puree. Bring to the boil and add tofu and blanched mushrooms.

3. Turn off the heat. Add evaporated milk. Season with salt and stir well. Serve.

ORANGE FROZEN YOGHURT CUP

INGREDIENTS

1/2 cup orange juice

1 1/2 cups plain yoghurt

1/3 cup evaporated milk

2 tbsp lemon juice

1 tsp vanilla essence

a pinch of salt

METHOD

Mix all ingredients together and divide among small serving cups. Keep in the freezer for 2 hours. Serve.

POINT

This frozen dessert may look rather humble, but it packs a refreshing citrus aroma beneath the creamy yoghurt.

CARROT CAKE

INGREDIENTS A

4 eggs

1 cup vegetable oil

1 cup sugar

1 tsp vanilla essence

INGREDIENTS B

2 cups flour

2 tsp baking powder

1 tsp baking soda

1/2 tsp salt

2 tsp ground cinnamon

INGREDIENTS C

1 carrot (peeled and shredded, about 2 cups)

METHOD

1. Beat the ingredients A together with a stand mixer over medium speed.

2. Mix ingredients B and sieve together. Add 1/3 of the mixture at a time to the egg mixture from step 1. Beat over low speed after each addition. Add ingredients C at last.

3. Pour the batter into a greased cake tin. Bake in a preheated oven at 175°C for 50 minutes. Let cool for 10 minutes before serving.

POINT

When I make this carrot cake in Canada, I'd beat a pack of Dream Whip (non-dairy whipped topping) with 1/2 cup of milk in a stand mixer over high speed for two minutes and top the carrot cake with that. You may also beat whipping cream until stiff and put a dollop on your carrot cake for equally great taste.

STEAMED CARAMEL CAKE

INGREDIENTS A
2 raw brown sugar slabs
(finely chopped)
1/2 cup boiling water

INGREDIENTS B
4 eggs
1/2 cup vegetable oil
2 tbsp honey

INGREDIENTS C
1 cup flour
2 tsp baking powder
1 tsp baking soda

METHOD
1. To make the syrup, put raw brown sugar slabs into a small pot. Add 1 tbsp of boiling water. Cook over low heat until golden. Add the remaining boiling water while stirring continuously. Let cool.
2. Beat the eggs over medium speed until well mixed. Add oil, honey and the cooled syrup from step 1 in that particular order. Beat until well incorporated.
3. Sieve all ingredients C together. Add 1/3 of the dry mixture to the egg and syrup mixture from step 2 at a time. Beat over low speed until well mixed after each addition. Pour into a greased cake tin. Cover with cling film. Let it rest for 1 hour.
4. Steam over high heat for 30 minutes. Serve.

POINT
1. I used raw brown sugar slabs in this recipe because of their caramel flavour and dark brown colour.
2. After pouring the batter into the cake tin, I covered it with cling film and let it rest for 1 hour. This step allows the batter to fully rise and the cake will be fluffier and lighter after steamed.

COCONUT POUND CAKE

INGREDIENTS A

225 g butter (at room temperature)

1/3 pack cream cheese (at room temperature)

1/2 cup sugar

3 eggs (at room temperature)

INGREDIENTS B

1 cup flour

1 tsp baking powder

1/2 tsp baking soda

a pinch of salt

INGREDIENT C

1/2 cup dried shredded coconut

METHOD

1. In a big mixing bowl, put in butter, cream cheese and sugar. Beat over high speed until well combined. Put in one egg at a time and beat well after each addition.

2. Sieve ingredients B together. Put 1/3 of the mixture into the cream cheese mixture from step 1 at a time. Beat over low speed to mix well after each addition. Stir in ingredient C at last.

3. Bake in a preheated oven at 175°C for 50 to 60 minutes. Serve.

POINT

1. You may also sprinkle with some of the dried shredded coconut on top. Just make sure you cover the cake with aluminium foil when baking. Otherwise, the dried coconut will brown too quickly and may even burn.

2. You may also use shredded candied ginger in place of dried shredded coconut. The steps are the same, but the cake tastes completely different.

JAPANESE SOUFFLÉ CHEESECAKE

INGREDIENTS A

250 g cream cheese (at room temperature)
40 g melted butter
6 egg yolks (at room temperature)
3 tbsp cake flour

INGREDIENTS B

4 tbsp whipping cream
2 tbsp plain yoghurt
2 tbsp milk
1 1/2 tbsp lemon juice

INGREDIENTS C

8 egg whites (at room temperature)
1/2 cup sugar

METHOD

1. In a mixing bowl, beat cream cheese and melted butter until well incorporated. Add one egg yolk at a time and beat till well mixed after each addition. Sieve in the cake flour.
2. In another mixing bowl, beat the ingredients B well. Add to the cream cheese mixture from step 1. Stir well.
3. In a third bowl, beat egg whites for about 10 minutes until stiff. Fold into the cream cheese mixture from step 2. Fold gently to mix well.
4. Pour the batter into parchment paper-lined cake tin. Bake in a preheated oven at 160°C for 1 hour. Serve.

POINT

Whether the soufflé cheesecake is fluffy and light depends very much on whether the egg whites are beaten till stiff. Yet, whether the egg yolks, cream cheese and butter are well combined. If you put in all egg yolks in one go, the batter will look like soybean dregs and the cheesecake will turn out stiff and rubbery.

ALMOND SPONGE CAKE

INGREDIENTS A

8 egg yolks

1/2 cup sugar

INGREDIENTS B

1 cup flour

1 1/2 tsp baking powder

1/2 cup ground almond

INGREDIENTS C

1/2 cup melted butter

INGREDIENTS D

8 egg whites

METHOD

1. Sieve ingredients B together. Beat ingredients A over high speed until well combined. Turn to low speed and add sieved ingredients B a little at a time. Fold to mix well. Add melted butter. Stir again.

2. Beat egg whites until stiff. Fold egg whites into the batter from step 1 a little at a time. Gently fold until well mixed.

3. Pour the batter into a greased cake tin.

4. Bake in a preheated oven at 180°C for 25 minutes.

CRISPY SESAME COOKIES

INGREDIENTS A

1/2 cup melted butter

2 tbsp sesame oil

3/4 cup sugar

1 egg

1/2 tsp vanilla essence

INGREDIENTS B

1 cup flour

1/2 tsp salt

1/2 tsp baking soda

3/4 cup sesames

METHOD

1. Mix ingredients A together. Add egg and vanilla essence. Stir again.

2. Mix ingredients B (except sesames) and sieve together. Add the dry ingredients a little at a time to the egg mixture from step 1. Stir well after each addition. Add sesames at last. Stir well.

3. Scoop out 1 tsp of dough and drop it on a greased cookie sheet. Leave some spaces around each cookie. Bake in a preheated oven at 170°C for 20 minutes. Serve.

POINT

Use white sugar and the cookies will taste less complex with a straightforward sweetness. On the other hand, brown sugar gives the cookies more depth and complex flavours.

謝謝您們
讓這食譜內容更豐富

香港浸會大學中醫碩士；香港中文大學中醫進修文憑；香港中文大學中醫骨傷科文憑
食譜內大部分湯水，都是由吳政栓中醫師提供。

吳政栓中醫師

鄧嬸

鄧嬸除了在溫哥華開設超級市場，更在電台主持街市行情報道，即席教授烹調方法，是家庭主婦的老友記！
她提供的蘿蔔丸子、鄧嬸芋頭糕吃過的人都讚不絕口。

老大 梅基偉

他提供的芋頭餅、碌雞，做法簡單又美味。

嘉幹

他是我的四弟，椰子蛋糕是他拿手的糕點。

何浪權

香酥芝麻餅是加拿大新時代電台同事提供的食譜，餅如其名，香酥美味。

歡迎加入 Forms Kitchen「滋味會」

登記成為「滋味會」會員
• 可收到最新的飲食資訊 •
• 書展 "驚喜電郵" 優惠 * •
• 可優先參與 Forms Kitchen 舉辦之烹飪分享會 •
• 每月均抽出十位幸運會員，可獲精選書籍或禮品 •
* 幸運會員將會收到驚喜電郵，於書展期間享有額外購書優惠

• 您喜歡哪類飲食叢書？(可選多於 1 項)
□ 中菜   □ 西菜   □ 點心   □ 烘焙   □ 湯飲   □ 甜品   □ 其他＿＿＿＿＿＿
• 您對哪類飲食題材感興趣，而坊間未有出版品提供，請說明：
＿＿＿＿＿＿＿＿＿＿＿＿＿＿＿＿＿＿＿＿＿＿＿＿＿＿＿＿＿＿＿＿＿＿＿＿＿
• 此書吸引您的原因是：(可選多於 1 項)
□ 興趣   □ 內容豐富   □ 封面吸引   □ 工作或生活需要
□ 作者因素   □ 價錢相宜   □ 其他＿＿＿＿＿＿＿＿＿＿＿＿＿＿＿＿
• 您如何獲得此書？
□ 書展   □ 報攤 / 便利店   □ 書店 (請列明：＿＿＿＿＿＿＿＿＿＿＿)
□ 朋友贈予   □ 購物贈品   □ 其他＿＿＿＿＿＿＿＿＿＿＿＿＿＿＿＿
• 您覺得此書的價格：
□ 偏高   □ 適中   □ 因為喜歡，價錢不拘
• 除食譜外，您喜歡閱讀哪類書籍？
□ 玄學   □ 小說   □ 家庭教育   □ 兒童文學   □ 語言學習   □ 商業創富
□ 兒童圖書   □ 旅遊   □ 美容 / 纖體   □ 現代文學   □ 消閒
□ 其他＿＿＿＿＿＿＿＿＿＿＿＿＿＿＿＿＿＿＿＿＿＿＿＿＿＿＿＿
• 您是否有興趣參加作者的烹飪分享活動？
□ 有興趣   □ 沒有興趣

姓名：＿＿＿＿＿＿＿＿＿＿＿＿ □ 男 / □ 女   □ 單身 / □ 已婚
職業：□ 文職   □ 主婦   □ 退休   □ 學生   □ 其他＿＿＿＿＿＿＿＿＿
年齡：□ 16 歲或以下   □ 17-25 歲   □ 26-40 歲   □ 41-55 歲   □ 56 歲或以上

聯絡電話：＿＿＿＿＿＿＿＿＿＿   電郵：＿＿＿＿＿＿＿＿＿＿＿＿＿

地址：＿＿＿＿＿＿＿＿＿＿＿＿＿＿＿＿＿＿＿＿＿＿＿＿＿＿＿＿＿＿＿＿

請填妥資料後可：
寄回：香港鰂魚涌英皇道 1065 號東達中心 1305 室「Forms Kitchen」收
或傳真至：(852) 2565 5539
或電郵至：info@wanlibk.com

＊請剔選以下適用的選擇
□ 我已閱讀並同意圓方出版社訂立的《私隱政策》聲明 #   □ 我希望定期收到新書資訊

食尋 。 蘊味

Savour the Homestyle Cooking

作者 | Author
黃淑儀 | Gigi Wong

策劃/編輯 | Project Editor
Catherine Tam

攝影 | Photographer
Imagine Union

美術統籌及設計 | Art Direction & Design
Amelia Loh

出版者 | Publisher
Forms Kitchen
香港鰂魚涌英皇道1065號 | Room 1305, Eastern Centre, 1065 King's Road,
東達中心1305室 | Quarry Bay, Hong Kong.
電話 | Tel: 2564 7511
傳真 | Fax: 2565 5539
電郵 | Email: info@wanlibk.com
網址 | Web Site: http://www.formspub.com
http://www.facebook.com/formspub

發行者 | Distributor
香港聯合書刊物流有限公司 | SUP Publishing Logistics (HK) Ltd.
香港新界大埔汀麗路36號 | 3/F., C&C Building, 36 Ting Lai Road,
中華商務印刷大廈3字樓 | Tai Po, N.T., Hong Kong
電話 | Tel: 2150 2100
傳真 | Fax: 2407 3062
電郵 | Email: info@suplogistics.com.hk

承印者 | Printer
合群（中國）印刷包裝有限公司 | Powerful (China) Printing & Packing Co., Ltd

出版日期 | Publishing Date
二〇一六年一月第一次印刷 | First print in January 2016

瀏覽網站　會員申請